Your job in the 80s

Your job in the 80s
a woman's guide to new technology

Ursula Huws
for Leeds TUCRIC

Pluto  **Press**

First published in Great Britain 1982 by Pluto Press Limited,
Unit 10 Spencer Court, 7 Chalcot Road, London NW1 8LH

Copyright © Ursula Huws 1982

ISBN 0 86104 365 0

Cover designed by John Finn
Cover illustrations by Jill Posener

Photoset and printed in Great Britain by
Photobooks (Bristol) Limited, Barton Manor, St Philips, Bristol

Contents

Dedication

To Jude and Pat, to show that your struggles have not been wasted

Acknowledgements

This book has been made possible by a research grant from the Equal Opportunities Commission to the Leeds Trade Union and Community Resource and Information Centre to study the impact of new technology on the working lives of women in West Yorkshire.

The two-year project drew heavily on resources in the TUCRIC library, on the knowledge and experience of a wide range of TUCRIC affiliates, on the academic and research skills of TUCRIC management committee members, and on the support and advice of the West Yorkshire Women and New Technology Group (WYWANT). Thanks are due to all of these.

A detailed account of the research project's first year has been published by Leeds TUCRIC under the title 'The impact of new technology on the working lives of women in West Yorkshire: interim report'. This includes a review of the literature, a discussion of methodology and terms of reference, statistical data and case studies, and is available from Leeds TUCRIC, 6 Blenheim Terrace, Leeds 2. A final report will be published in due course by the Equal Opportunities Commission.

Every attempt has been made to contact the holders of copyright in the illustrations reproduced in this book, but in a few cases this has not proved possible. The author would be glad to hear from any individuals claiming copyright in any of this material.

Introduction

In 1978, BBC television showed a Horizon film called *Now the Chips Are Down* and, almost overnight, new technology became a national talking-point. The film showed how tiny chips of silicon were taking over from the cumbersome mainframe computers of the past, and making it possible to use computing techniques simply and cheaply in a wide variety of applications. Most dramatically, these 'microprocessors' were shown in action in all kinds of workplaces – taking over many of the functions of storekeepers, stock-clerks, typists, paint-sprayers and junior doctors.

Suddenly, the question on everyone's lips was, 'How many jobs is new technology going to destroy?' The answers were many and varied. Some experts predicted unemployment levels of up to five million in Britain by the end of the 1980s. Others said that new technology would actually create more jobs than it eliminated.

Several commentators drew particular attention to women's jobs, but here too their predictions were contradictory. Some said that new technology would open up new jobs for women by removing the skill and strength requirements from traditionally 'male' occupations. Others believed that large numbers of women would be put out of work by the introduction of microelectronics.

One of the things that makes it hard to work out what the effects of new technology will be is the fact that it is not introduced out of the blue, into workplaces in which nothing else is happening. There are always other factors at work as well, influencing the number and quality of jobs, and it is often very difficult to pinpoint the exact reason why a job has disappeared. The years since 1978 have been a period of recession, when unemployment has risen steeply for a variety of reasons, only some of them directly connected with new technology.

In private firms, a sharp fall in profits and increasing international competition has caused the closure of many factories and the 'rationalisation' of others. Some, but not all, of this 'rationalisation' involves investing in new, job-destroying machinery.

In government and other public employment, the crisis has meant massive cutbacks in expenditure – which also lead to redundancies and 'rationalisation'. Again this is sometimes achieved by bringing in new labour-saving technology.

Trying to develop strategies to protect jobs in the face of the recession, workers and their representatives have found themselves with a bewildering array of new problems to contend with, ranging from changes in employment laws to regional development policies, from short-time working schemes to public expenditure cuts. New technology has often been seen as just one 'issue' among many, and often less pressing and immediate than others.

Women, in particular, have felt the force of the recession in their home lives as well as at work, as deteriorating public services and worsening living standards have taken their toll. They have had little space to step back from the daily grind and take a longer term view of their lives and the transformations which new technology might bring. Most are vaguely aware that it could bring major changes, but it is difficult to connect the sweeping, space-age visions of the television pundits with the humdrum details of one's own everyday existence, and there's usually a more urgent problem screaming for attention, so few have been able to develop strategies for dealing with it in advance.

Here is a description of her initial response to new technology by Pat McDougall, convenor at a vehicle components factory in Calderdale:

> The effect new technology had on me initially was one of total panic. Automation and its effects were something I had some awareness of, but here we were, a factory full of semi-skilled women workers faced with the ultimate automation. Up till then we had some power, not much but some: as long as the

employers needed to buy our labour we had something to bargain with. When they did not need it we were powerless.

We felt helpless in the face of a development which not only would crush us but also, at the time, seemed to make economic sense. All the arguments were used: 'We must get into the race or we will be losers.' 'You cannot hold back progress.' 'We don't want any latter-day Luddites here.' And that was just from the union side.

The management played its significance down, and tried to project it as an event which would take place in the far distant future – we shouldn't worry our little heads about it. They used this on the members very successfully.

When you are a woman, working all day, working at home, trying just to keep the family together, smoothing out everyone else's problems, it's difficult to worry about something so far in the future. 'Please God, just let me make it through till bedtime' is about all you can hope for. When a male management comes along and reassures you that everything will be all right, you want to believe it, even though you know deep down it's not so.

In fact, Pat McDougall showed considerable foresight compared with many people in similar situations. Often, it is not until some time after the new technology has been introduced into a workplace that the full implications of the change become apparent. And by then it is frequently too late to avoid many of the worst consequences.

So, what are the effects of new technology? Does it improve the quality of work, and take the drudgery out of it – or does it make jobs more boring and repetitive? Will it lead to new opportunities for women – or will it simply reduce the number and variety of jobs open to them?

These were some of the questions that led to the setting up, in 1979, of a two-year research project at Leeds TUCRIC, funded by the Equal Opportunities Commission, to investigate the impact of new technology on the working lives of women in West Yorkshire.

TUCRIC, the Leeds Trade Union and Community Resource and Information Centre, was uniquely placed to carry out the project, since it was able to combine a research facility with extensive contacts among trade unions, community groups and women's groups in the area, by and for whom it is run. Previous work by the centre had uncovered an urgent need for information on this subject, and had also revealed that, in women's employment, West Yorkshire was a fairly typical area, so that lessons could be drawn for the rest of Britain from the project's findings. The characteristic pattern of women's employment was one where, traditionally, manufacturing industry had played a large part, but this had been declining for some years. Up until the mid-1970s, new jobs in expanding services had made up for the loss in manufacturing industry, although many of these tended to be insecure and part-time. With the recession, however, the services had stopped growing and women's unemployment was on the increase. The only major difference between West Yorkshire and the national picture was that in West Yorkshire the situation was slightly exaggerated, with more manufacturing jobs having disappeared and fewer service jobs being created. However the picture was much nearer the average than that of London and the prosperous South East on the one hand, or, on the other, depressed industrial areas like South Wales, Tyneside or parts of Scotland.

The first task was to find out what jobs women were actually doing, and how new technology might affect them. Another part of the project's work involved a survey of workplaces in the area. Representatives from about forty workplaces filled in detailed questionnaires, giving information about: how new technology had been introduced where they worked; the response of management, unions and ordinary workers; changes in grading; promotion prospects; skill levels; health and safety and a number of other factors. This is the survey meant by the 'TUCRIC survey' in the text of this book. A detailed analysis of the results will be published by the Equal Opportunities Commission, Overseas House, Quay Street, Manchester 3, for those interested in knowing more about the survey and the rest of the project's work.

The purpose of this book is to spread the project's findings to a wider audience, in particular to working women and to those concerned about their welfare and future prospects. In the process, I hope that it will also make a contribution to the debate currently being conducted in the labour movement, in women's groups and elsewhere about what our response should be to the introduction of new technology.

The book has been designed so that busy working women

who may not have time to read it from cover to cover can use it as a guide and reference book. Look at the contents page to find out which chapter is relevant for your kind of work, and read that first. You may find enough information there to answer your questions, or tell you where to go for further information. If not, turn to the chapter on negotiating new technology at the end of the book. This contains a checklist of points to cover in negotiations over the introduction of the technology, and suggestions for where to find out more.

1

Where Do Women Work?
The Statistical Background

Although four out of ten workers in Britain now are women, they are not spread evenly throughout the 26 industries which make up the official picture of the British economy, as men are. On the contrary, as this chart shows, women are heavily concentrated in relatively few types of work.

The 11 industries shown here are those in which over 150,000 women were working in June 1981. The top six of these industries, with by far the largest numbers of women workers, are all service industries, with five manufacturing industries bringing up the rear.

The biggest of all is professional and scientific services, employing over two-and-a-half million women, compared with fewer than half that number of men. Most of these workers are in schools and hospitals.

The second largest industry, with a million-and-a-half women, is distributive trades, which includes shops, warehouses and some offices.

Miscellaneous services comes next, covering a rag-bag of different types of workplaces ranging from hairdressers to petrol stations, from estate agents to restaurants. Insurance, banking and finance and public administration and defence, as their names imply, are both office-based industries, employing women in banks and in the offices of insurance companies and government departments.

Transport and communication also employs some office workers, but the bulk of the workforce in this industry is employed on trains, buses and so on. Most of these workers are men – there are

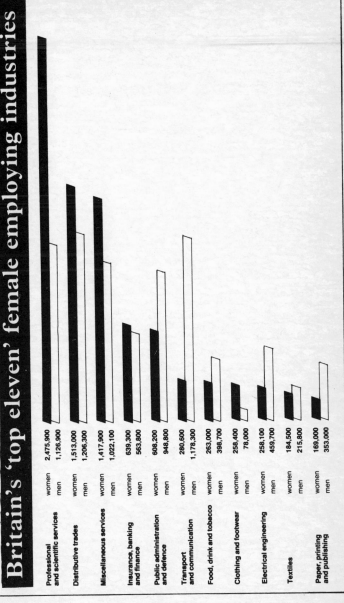

Britain's 'top eleven' female employing industries

Professional and scientific services	women	2,475,900
	men	1,126,900
Distributive trades	women	1,513,000
	men	1,206,300
Miscellaneous services	women	1,417,900
	men	1,022,100
Insurance, banking and finance	women	639,300
	men	563,800
Public administration and defence	women	608,200
	men	948,800
Transport and communication	women	280,600
	men	1,178,300
Food, drink and tobacco	women	263,000
	men	398,700
Clothing and footwear	women	258,400
	men	78,000
Electrical engineering	women	258,100
	men	459,700
Textiles	women	184,500
	men	215,800
Paper, printing and publishing	women	169,000
	men	353,000

nearly four times as many men as women in this sector – but there are nevertheless large numbers of women, especially in telecommunications and in the office-based jobs in transport.

Of the five manufacturing industries with large numbers of women, three come as no surprise: food and drink, clothing and footwear and textiles are all industries closely connected with the work traditionally done by women in their own homes. Of the other two, electrical engineering is a comparatively new industry, which grew up in the post-war years when many women were looking for work. The repetitive assembly work involved was considered ideal for women, whom the employers considered to have greater patience and dexterity than men, as well as being less likely to demand high wages. Paper, printing and publishing is more than two-thirds male as an industry. Where women are employed, they tend to be concentrated in the lower-paid, lower-grade jobs such as packing.

Most of the statistics available, from the Department of Employment and elsewhere, classify work according to industry, as this chart does. While they give us some useful information about what types of employers people work for, they don't tell us what sorts of jobs they are doing – the most important information for working out how employment is likely to change in the future.

Many jobs, for instance cleaning or clerical work, are to be found in all of these industries – from engineering factories to hospitals, from coal mines to shops. If you are looking for a job as a typist, say, or a canteen assistant, you are not going to care whether your employer is categorised as 'distributive trades' or as 'paper, printing and publishing'. You will be much more interested in things like what the pay is, what the hours are, and how easy it is to get to.

And if a new machine comes along, which affects the work of a typist or a canteen assistant, its effects will similarly spread across a lot of different industries, and affect the job prospects of all women in that occupation.

So, to begin to work out the likely effects of new technology on women's employment, we need to know, not who their employers are, but what their occupations are. Unfortunately, this information is only collected every ten years, in the census of population. Since the latest figures available at the time of writing are those of the 1971 census – far too out-of-date to be really useful –

100 women

clerical workers

factory workers

'housework' workers
cleaning, cooking etc

'caring' workers
teaching, nursing, social work etc

sales and distribution workers

other employment

school age or below

over retirement age

registered unemployed

full-time education or training

disabled

non-registered unemployed,
homeworkers and full-time housewives

based on statistics for entire female population of West Yorkshire

it has been necessary to do some fairly elaborate arithmetic, comparing the census figures with the industry-based ones for the same year, and then extrapolating from those to more recent industry-based figures, to come up with an estimate of what jobs women do, and where. This exercise was carried out on the figures for West Yorkshire, and the chart on page 15 is a summary of these figures, rather than the national ones. However the West Yorkshire picture is very similar to that for the rest of the country as a whole. The only major difference is that slightly more women work in factories, and slightly fewer in offices, than in the rest of Britain.

The chart shows the entire female population, what proportion of women work, and the sorts of jobs they do: of every 100 women, 24 are at school or over retirement age, 2 are in full-time education or training, 2 are disabled, 3 are registered unemployed, 37 are non-registered unemployed, homeworkers or full-time housewives. The rest, 32 per cent, are waged workers.

As you can see from the chart, most of the 32 women out of every 100 who work are concentrated in an astonishingly small number of occupation types.

By far the largest group – nearly one-third – are clerical workers. The second largest, accounting for eight out of 32, or a quarter of all working women, are factory workers. The types of work these women do are more diverse than those of the clerical workers and the group includes assemblers; packers; textile trade jobs such as winders, reelers, spinners, twisters and weavers; sewing machinists; electrical and electronic assemblers; food processing workers; and other factory work, mainly unskilled or semi-skilled.

The third category, covering 7 out of the 32 working women, is a category which we call 'other people's housework'. It includes cleaners; canteen assistants, counter-hands, cooks, kitchen hands, general servants, waitresses and launderers.

The fourth category covers the professional 'caring' jobs (you could say that these too are extensions of the housework role) and comprises nurses, teachers, social workers, nursery nurses and a few other 'professional' jobs, such as physiotherapy. This accounts for 4 out of the 32, or one-eighth of working women.

The only other type of work where any significant number of women is employed is sales and distribution. This includes some packers and labellers, storekeepers, warehouse assistants and some proprietors and managers as well as cashiers and sales assistants. These account for 3 out of 32 or roughly a tenth of working women.

Although there are, obviously, a large number of other occupations in which women can be found, the numbers are fairly low, and not likely, in themselves, to have a great impact on opportunities for the majority of women. Taken all together, they account for only one per cent of all women, less than 1 in 32 of working women.

In the rest of this book, we concentrate on the 'top five' types of occupation, since the effects of new technology on these will affect the biggest numbers of women in this country.

2

Clerical Work

Once upon a time, an office job was considered one of the best a woman could hope to get. If you dressed carefully, watched your language and were reasonably attractive; if you went to secretarial college and worked hard for a year or two at your shorthand and typing and got up to good speeds; then, so the belief went, you might find yourself a job that was secure, clean, friendly, varied and unpressured, with regular hours, paid holidays, regular pay increases and – if that was what you wanted – even a good chance of picking yourself up a well-heeled husband at the Christmas party.

Although anyone who had slaved in a large typing-pool could tell you otherwise, there was just enough truth in this idea to make it very difficult to contradict, particularly since it was reinforced in the media by an image of the female office-worker as a combination of sex-object, status-symbol and housewife.

In fact, very few women have ever achieved the status and freedom of being a 'personal secretary'. For the vast majority office work has always been fairly boring and alienating, though it would be fair to say that, at least for the first part of this century, it was better than factory work, shop work or domestic service (the only alternative jobs for most working-class women).

However, office work has changed considerably over the past few decades. For one thing, there are simply many more office workers than there used to be. In 1850, white-collar workers were only one per cent of the working population, a proportion that had risen to 10 per cent by 1950, and to a staggering 40 per cent of the workforce by 1970.

There were several reasons for this rapid growth:

The post-war boom brought a general expansion of the economy, with increases in banking and insurance, new industries growing up, and more and more sophisticated ways of marketing consumer goods, all of which created new white-collar jobs.

The growth of the welfare state, too, meant a massive expansion of administrative work in government agencies ranging from hospitals to planning departments, from tax offices to universities.

Finally, the increasing automation of factory work brought about a need for more 'scientific' management, and a shift from shop-floor to white-collar jobs in manufacturing industry, as managements found that they had to keep track of more and more records such as how much was being produced and how fast, or which workers had earned how much in productivity bonuses.

In short, the post-war years saw a huge increase in the *amount of information* to be collected and sorted and stored and then retrieved, in every type of institution. A new army of workers was needed to handle all this information. Where was it to be found? In the booming post-war period there were few unemployed men to take up these new jobs – in fact immigrants were being encouraged to come to the industrialised countries to do the more unpleasant jobs which men who had fought in the second world war were no longer willing to do. There was only one place this new workforce could come from – the home. The vast majority of the new army of office workers were women, and a very high proportion of them were married.

As offices got bigger, the work was broken down into more and more specialised functions, making it more boring and repetitive for individual workers. During the same period, the status of office workers began to fall, as did their wages in relation to factory workers. It was also during this period that, for the first time, white-collar workers began to join trade unions in large numbers.

Managements were beginning to worry about the 'productivity' of their office staff, just as if they were factory workers. There were so many office workers now that their wages formed an increasingly high proportion of the total bill for labour and articles began to appear in the trade press, examining ways in which office

The office wife

I could be just your cup of tea

I'm just waiting to give you details of the ideal low cost drinks service.

Drinkmaster offers a choice of 17 different drinks. Each one made from sealed capsules to ensure fresh drinks every time. Plus a range of attractive dispensers that just plug into electricity supply. Phone today I am longing to hear from you.

Popular images of clerical workers, like these cartoons, invariably show them servicing their bosses not as fellow-workers, but as men. Whether she's a sex-pot turning him on and gratifying his ego, a mother-figure looking after his health, or a servant bringing him cups of tea, the office worker's role is seen as almost identical to that of a housewife, never as what she really is—a skilled worker getting on with her job.

Early computers

In the 1950s and 60s, computers were still huge mysterious machines. Cumbersome and erratic, they had to be kept in special environments and tended by skilled, relatively high-status male operators and technicians. The expense involved meant that their use was limited to large institutions like banks, universities and government departments.

Picture by courtesy of Ikon

overheads could be reduced. Meanwhile, of course, the tide of information kept on rising.

The first computers made their appearance in offices in the 1950s. They were huge, cumbersome machines, requiring a lot of space and specialist treatment – they had to be kept at particular temperatures in dust-free environments, (something a lot of office workers would have welcomed for themselves!) and they were enormously expensive. Only very big-scale operations, involving a lot of difficult calculation, could justify the expense of one of the early 'mainframe' computers, as they were called, and their use was limited to very big organisations with a lot of financial transactions to make, or specialist mathematical research to carry out – banks, government departments, the head offices of multinational corporations and big military establishments. The only sort of information generally handled by these machines was numerical. Although they could type out words on their noisy printers, they weren't used for 'processing' them, as they were for numbers.

People got used to seeing computerised bills, pay-slips and bank statements, but the vast majority of office workers was untouched by the introduction of the machines – they went on typing the letters and doing the filing very much as they had always done – except that often the rewards weren't so good and the supervision was stricter. And because the amount of office work was expanding so fast, nobody noticed any jobs disappearing as a result of computers being brought in.

During the next twenty years or so, partly as a result of heavy investment in research by the US military and space authorities, computer technology developed apace. The actual functions a computer could carry out didn't change much, but the machinery that carried out these functions became much smaller and cheaper as, progressively, clumsy, fragile, overheating and unreliable valves were replaced by transistors, which, in turn, were ousted by silicon chips. This process of miniaturisation did not just make machines smaller and cheaper; it also made them much less delicate and more portable, so they could be put into a whole variety of different appliances, in widely varying environments, without coming to any harm. As applications for these new computers increased, and sales grew, it became feasible to mass-produce them, using cheap third world labour, thus bringing the prices down still further.

By the mid-1970s, the stage was set for major changes in office work. The boom days of the fifties and sixties were over, profits were down and few industries were expanding. Employers were getting very worried indeed about the costs of running their offices. Meanwhile, as life became more intricate, the amount of information was still growing. Some office workers had organised in unions and, exposed to the ideas of the women's liberation movement and the 'new trade unionism' of the late sixties and early seventies, were beginning to demand decent wages and conditions. This worried employers even more. How could they reduce their costs, gain some control over a workforce that was beginning to seem a bit stroppy, and make the work simpler to do, so they would not be forced to limit their recruitment to a small group of highly-skilled workers?

The answer was: automate. And the technology now existed to computerise a whole range of information-processing operations simply and cheaply, without the need to provide special environments. Computers need no longer be restricted to processing numbers, but could now be used for words, designs, photographs, and even speech. Television-style screens, known as visual display units (VDUs) could be used for putting information into and taking it out of the new computer systems at the touch of a button, while whole filing-cabinetsful of information could be economically stored on cassettes or 'floppy discs'. The computing itself – the 'thinking' work of the machine – was done by minute microprocessors, so tiny that several could be fitted inside a wristwatch.

Soon manufacturers were producing equipment specially designed to carry out office information-processing functions. Perhaps the best known of these machines is the word processor, which can store text in its memory, make corrections automatically, and vastly increase the productivity of a typist. Attached to a printer, it can produce traditional-style paper documents; attached to a central computer, or linked across GPO wires to other machines in other offices, it can bring about even more startling changes in office work, by eliminating paper altogether and creating a situation where, for many routine types of communication, one machine can talk directly to another, cutting out human intermediaries altogether. Other types of machines automate the functions of filing clerks, telephonists, drafters and other office workers.

The machines are so cheap that most cost less than a year's wages for one of the workers they displace, although some require an expensive infrastructure – e.g. new cable connections between offices, or a specially programmed central computer – in order to reach their full effectiveness.

What does new technology mean for office workers?

The prospect of working with new technology is one which most people welcome when it is first suggested. (In the TUCRIC survey, there was only one workplace where workers said they were 'opposed to new technology' before it came in.) The reasons for this are understandable enough. Its introduction is often combined with a general modernisation or re-design of the workplace, which can make a huge difference, especially to workers who may have been asking for years for new chairs, or carpets, venetian blinds or a different lighting system. Not only does it bring with it the promise of an improved and up-to-date working environment, but it also seems to offer a chance to learn new skills and get rid of some of the more boring and repetitive aspects of many jobs. In addition, many managements offer a financial incentive for accepting new technology – in the form of extra payment for working the new machines, or a rise on completion of a training course in how to use them. Some workers, too, see the new machines primarily as toys at first, and look forward to the chance to play with them.

Some of these advantages are real and lasting; others tend to disappear with the daily reality of working the machines. They are, however, fairly well known – which cannot be said of most of the disadvantages the new technology can bring. What are these disadvantages?

Job loss

Perhaps the best known effect of new technology on office workers, and the one which causes the most public concern, is the loss of jobs. Predictions of how many jobs are likely to disappear vary widely. Some experts think that as many as four out of ten office jobs will disappear by the end of the 1980s as a result of the new technology, while others forecast only one out of ten, or even less.

Miniaturisation

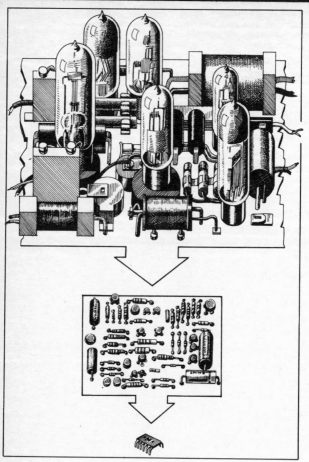

The silicon chip is basically a very small electronic circuit, a tiny replacement for transistors which, in their turn, replaced the valves from which early electronic circuits were made. Here you can see the difference in size between valves, transistors and chips. The capacity of the circuitry on a chip is much higher than that of its predecessors'. The entire circuitry of a room-size early computer can now easily be fitted onto a chip less than a centimetre square.

Picture by courtesy of the TUC

Twenty-five to thirty per cent is a more typical estimate. Whatever the exact figure is, it is clear that job loss can be very drastic when new machines are brought in.

In one mail order firm in West Yorkshire, for instance, full-time clerical staff were reduced from 1,000 to 550 when the firm was computerised, and part-timers were cut by half, from 100 to 50.

In some cases, job loss is more gradual, achieved over a period of months by non-replacement of staff who leave. In yet others, where the amount of work is expanding, there may be no jobs lost at all. For instance, the Halifax Building Society did not make anyone redundant when they progressed from automatic typewriters to word processors – but they managed to get three times as much work done with the same staff.

However it is always as well to remember, when new technology is being brought in, that the main reason for introducing it is to increase productivity: in other words to get more work done by fewer people. At a time when few private industries are expanding and most public ones are cutting back, it is reasonable to expect that the emphasis is more likely to fall on the 'fewer people' than the 'more work' – and job loss, sudden or gradual, is to be expected.

'One computing survey, quoted by the white-collar union APEX in 1980, 'revealed that more than two-thirds of private sector employers planned to increase the money invested in computing, despite the rapid drop in costs of equipment. A primary reason was to cut labour costs. Companies estimated that for every £1 spent on data processing alone, £2.60 was saved in clerical and other costs.'

from *Automation and the Office Worker*, APEX, 1980

De-skilling

In the past, clerical work involved a range of skills, from shorthand to accurate typing, from knowing how to lay out a variety of documents to operating such office machinery as duplicators, franking machines and addressographs. A secretary was expected to be able to run an office in her boss's absence, so secretarial colleges insisted that their students learn about all aspects of office work, from dealing with outside agencies like the Post Office and the Department of Employment to correcting the boss's spelling mistakes.

'The problems which are usually associated with office automation . . . make up the final chapter of a story which began with the industrial revolution, that is with the supreme assertion of capitalist production over all preceding types of production . . . The taylorisation of the first factories . . . enabled the labour force to be controlled and was the necessary pre-requisite to the subsequent mechanisation and automation of the productive process . . . Information technology is basically a technology of coordination and control of the labour force, the white collar workers taylorian organisation does not cover.'

Franco di Benedetti, Managing Director of Olivetti, at a *Financial Times* conference in 1979

Not every office worker had a chance to use all of these skills, but even if she was stuck in a typing pool all day, or spent all her time filing, she had a fair degree of control over her own work compared with a factory worker, and often needed to use her personal judgement and skill to work out the best way to do a job. Usually, for instance, she could decide in what order to do various tasks, how fast to do them, and when to take a break.

New technology brings drastic changes in skill requirements. A word processor operator no longer needs to know how to centre, tabulate or lay out documents. Nor does she need to know how to insert carbons or make corrections or when to start a new page. All these, and more, the machine can do for itself. It may also have a number of standard paragraphs and phrases stored in its memory, so all she must do is key in details, such as the name and address of a customer, or the date of a contract. Some machines will even automatically correct her spelling mistakes! Managements, then, do not any longer have to recruit workers who have good educational qualifications and therefore expect a decent wage. They can go for the cheapest workers they can find, who can be given the basic training to become reasonably fast keyboard operators.

For the workers, it means loss of the job satisfaction that comes from exercising your skill and knowing that you are doing a job well, loss of control over the work, and increased boredom.

In the TUCRIC survey, representatives of 20 workplaces answered the question, 'Do you think that workers have more or less control over their work since the new technology was introduced?' Of these, three-fifths (representing 12 workplaces) felt there was less control than before, while three felt there was no change. Of the five who felt there was more control than before, none was describing a typical clerical job – they were managers, technicians or non-clerical posts such as research assistants.

They were also asked if they had noticed any other changes as a result of the new technology. Here are some typical responses: 'Boredom'. 'Management are at us more often now to clear the work quicker which means more work and staring at screens.' 'Workers need to think less, though concentrate more.'

Loss of promotion prospects

This de-skilling also has the effect of making office work less varied. Instead of doing a little bit of a lot of different jobs, as many office workers did in the past, there is a tendency for office workers to spend all their time doing the same thing. That way, management can be sure that their expensive new machines are not sitting idle for part of the day. The work process becomes more and more fragmented, with different people doing different parts of it, instead of one person seeing a whole process through from beginning to end. Workers do not have a chance to learn about the whole process any more, but only know about their own little bit of it, like many factory workers. And without knowledge of the whole process, the chance of learning enough to be promoted becomes very slim.

It also seems likely that there will be fewer positions to be promoted to. Many of the new machines have built-in monitoring devices, which keep tabs on how fast each worker is doing her work. This means less need for supervisors – traditionally one of the ways up the career ladder for clerical workers. In modern offices, the better jobs tend increasingly to be technical ones, connected with computer maintenance, programming or systems design – jobs for which few clerical workers are qualified.

In the TUCRIC survey, 22 people answered the question on promotion prospects. Of these, 13 said that they had deteriorated

A word processor

A word processor operator at her 'work station'. As is often the case, this word processor has been introduced into a conventional office, with no special provisions made for its user. In a typical application, she has to move her eyes backwards and forwards between printed material (known as 'hard copy') and the screen, under lighting conditions designed for traditional types of clerical work, which are taking place alongside. Gone are the separate computer rooms of the 50s and 60s, and with them, any privileged treatment for computer operators.

Picture: Sefton Photo Library

since the new technology was introduced, with only seven reporting no change, and only two (neither of them clerical workers) considering that there had been an improvement. This seemed to be a question people felt strongly about, with comments like 'Gone for good!' written on the questionnaire.

Stress

Another result of this increasing specialisation and speeding up of office work is a marked increase in stress. Thirteen out of 15 of the workplaces in the TUCRIC survey reported more stress after new technology was brought in.

Several factors contribute to this marked increase in stress. There is extra pressure on people as a result of the increased speed and monotony of their work – the need for continual concentration, without the ups and downs in concentration levels of more traditional work. This may be accentuated by dissatisfaction as a result of feeling that one's skills are wasted and that the job no longer has any interest or satisfaction.

Stress can also be directly caused by the physical conditions of working with much of the new machinery. Most work with computerised equipment involves either feeding information into, or retrieving information from the machine, usually via a television-style VDU screen, and many jobs involve spending most or all of a working day staring with concentrated attention at one of these screens. This puts considerable strain on the eye muscles. The human eye is a complex and delicate mechanism, with six separate sets of muscles designed to allow you to look near and far, up and down, at brightly or dimly lit objects, in a variety of directions. In most normal circumstances, it has a chance to do all of these things over a period of an hour or two, but is not required to do any one of them for long periods.

Work with a VDU screen requires the eye muscles to do the same thing for very long periods. Like any other muscle in your body (imagine having to stand on one leg or hang by one arm, or keep your toes curled up for seven-and-a-half hours a day,) they come under enormous strain as a result, a strain that generates the symptoms of stress in other parts of your body: loss of appetite, headache, dizziness, indigestion, insomnia, depression, irritability, anger and nervousness. Here is a typical comment, from one of the workers in the TUCRIC survey working at a screen all day:

If I had known about it before as I do now I would advise other people to look for another job. We are on it 6¾ hours a day and we're shattered by hometime. We are all fed up with it but other jobs are not so easy to find at the moment and as our office has nothing else but computers there is no change during the day to rest our eyes from it.

For many women, of course, stress is not something that is confined to the workplace. Many, particularly working mothers, are over-burdened with responsibilities and exhausted before they even set foot in the workplace. Rising unemployment, increasing insecurity and cutbacks in services have added to these pressures, so that many women are already chronically stressed, and may be taking tranquillisers, anti-depressants or pain-killers for the symptoms. Some will not even think of the effects of their work on their health, and will go on blaming themselves or their family situation, not realising how much of their stress is caused by work conditions. For them, the problem is even more serious. Tests have shown that librium, valium and alcohol can slow down some of your eye muscles, making them work less efficiently and making the stress even worse.

Other health effects

The introduction of new technology can bring with it a number of other health hazards. Some are fairly obvious – such as the dangers from trailing leads and from the overheating caused by some machines. Others are the result of poorly designed equipment and furniture – such as backache from incorrect screen angles and badly designed chairs.

In the TUCRIC survey, we asked respondents to describe some of the health effects of the new technology. The results speak for themselves:

Out of the 13 workplaces whose representatives filled in this section.

All	singled out an increase in	**Headaches**
Ten	singled out an increase in	**Eye trouble**
Nine	singled out an increase in	**Tiredness**
Five	singled out an increase in	**Backaches**
Three	singled out an increase in	**Depression**

Three	singled out an increase in	**Consumption of tea and coffee**
Two	singled out an increase in	**Wrist pain (tenosynovitis)**
One	singled out an increase in	**Use of tranquillisers**
One	singled out an increase in	**Consumption of alcohol**

Other health effects mentioned included neck ache, red eyes, deteriorating eyesight, sleeplessness, eye strain, elbow pain and restlessness.

Another serious health effect, which TUCRIC has come across although it did not occur in any of the workplaces in the survey, is the triggering of epileptic fits, which can sometimes be caused by the flicker of a VDU screen.

Changes in working hours

There is another, more insidious effect of new technology, less often noticed, but with serious implications for women in the long run: a tendency to a lengthening of the working day and the introduction of shift working.

If you stop to think about it from management's point of view, the reasons for this development become clear. To give an over-simplified example, it is obviously going to be cheaper to buy two machines and have them working for sixteen hours a day, than it is to buy four to be worked for eight hours a day each. The extra wear and tear on the machines can usually be discounted when the machines are, in any case, becoming obsolete very rapidly, and costs are still tumbling.

It is therefore going to be in the management's interests to have the machines worked for as long as possible, to get their money's worth out of them. It is unlikely that many organisations will change overnight to full, round-the-clock factory-style shift working. What seems more likely to happen – and there is some evidence that it is already happening – is that there will be a gradual extension of normal working hours, not necessarily involving the existing nine-to-five workers.

One example of this is the introduction into offices of the evening 'twilight' or 'mum's' shift, already a familiar feature in areas where there are factories employing large numbers of women. Women with young children, who cannot get childcare during the day, are often desperate enough for work to take on these 'twi-light' jobs at wages and conditions well below those for daytime workers.

The extent to which offices are moving over to this type of working is difficult to measure, but it is certainly becoming more common, particularly in large institutions like insurance companies and building societies with a lot of inter-branch business. An additional reason for operating after six and at weekends for many offices is that at these times they can take advantage of cheap GPO rates for off-peak communications. As electronic communications between computers in different offices become more common, this will become increasingly advantageous. Small organisations which buy computer time on big specialised computers belonging to large institutions like universities are also likely candidates for unsocial hours working, since they too will wish to take advantage of cheap off-peak rates.

In fact 24-hour working has been associated with computers ever since they were first introduced, and many workers in large offices got used to thinking of the people who worked in the data processing department as a race apart. Not only were they almost invariably male in a largely female environment, but the fact that they worked shifts often meant that they came and went at different times from everyone else, were paid differently, and altogether seemed bound by a different set of rules.

As computers become simpler to operate and cheaper, much of the skill requirement and mystique of the computer operator disappears, and with it much of the special treatment data processing staff have become used to. Some high-grade design and programming jobs will remain, but run-of-the-mill data processing workers increasingly find themselves sharing the work they do with other office workers. They are beginning to lose their advantages, but often find themselves stuck with the *disadvantage* of having to work nights, with all its attendant physical and emotional problems. And this too, is beginning to spread to other office workers.

In banks, for instance, the introduction of new technology for round-the-clock cash dispensing machines has meant 24-hour working for all the back-up staff needed to keep the system functioning.

The introduction of shift working has serious implications for women. In offices they are not actually banned from working

Computing in the home

Packaged more as an executive's toy than a serious working tool, this 'teleputer' combines videotex and local computing with a video cassette recorder. According to its promoters it can be used for, among other things, electronic games; booking holidays, tickets or theatre reservations; sending and retrieving electronic mail; watching or recording TV programmes or video discs; learning a language or a skill like carpentry; home-based shopping or 'checking current stock exchange prices and sending an electronic message to your brokers'. Many experts predict that machines like these will also be used for people to work from their homes, but how much fun would it be to be tied to a screen like this all day for your living, without the companionship of other workers?

Picture by courtesy of Rediffusion Computers Ltd

nights, as they are in factories, but domestic responsibilities make it impossible for most women to take on night work. If willingness to work nights becomes a requirement for large numbers of office jobs, then it will effectively keep women out of those jobs and create a form of discrimination formerly only found in factories.

The Hazards of Shiftwork

Shift work is not good for you if you are young, old, married, a parent, socially or politically active, unhappily married, Moslem; or if you like to equally share the responsibility of children, to eat with the family, and to have regular sex with your partner; or if you live in inadequate or overcrowded housing, have a large family, have poor sound-proofing in your home, have inferior canteen facilities on the night shift, have a boring repetitive job; or if you suffer from diabetes, epilepsy, gastro-intestinal problems, duodenal ulcers, colitis, coronary thrombosis, or stress. Shiftworkers who find themselves described above are likely to be amongst the majority of shiftworkers who don't like shiftwork. They may also be amongst the 20 per cent of shiftworkers who fail to 'adapt' and who disappear from shift work altogether, taking with them their higher rates of mortality, morbidity, psycho-neurotic disorders, social and marital problems, plus the lower wages of normal hours.

Some of the findings in a review of the evidence on *Shift Work and Health* by Dr J. M. Harrington of the TUC Centenary Institute of Occupational Health, published by the Health and Safety Executive.

New types of homeworking

Much of the publicity surrounding the introduction of information technology has centred on images of new technology in the home. In the future, we are told (they don't say exactly when), offices will no longer be necessary and we will all have an array of gleaming new machines in our living rooms from which we can do anything from ordering our dinner to arranging our holidays; from paying our electricity bills to watching a vintage movie. The home is also where we will work. The picture is an alluring one – the mocked-up 'home of the future' in the studio is luxuriously designed and comfortable-looking and the sorts of work being carried out are invariably those associated with the lifestyle of a successful middle-class executive.

It's a big jump from this image to the reality of the conditions in which most homeworkers work today.

Homework, typically, is something nobody does from choice; it is a last resort for people – almost invariably women – with no other source of income who are trapped all day in their homes by disablement, by language difficulties, or by the need to care for young children or for old or handicapped dependents.

Research has shown that homeworkers are the most disadvantaged group of workers, with lower wages and fewer benefits than any other group (with the possible exception of prisoners). In addition, they are often exposed to health hazards and frequently suffer acutely from their isolation.

Since no research has been done, nobody knows the extent to which new technology is leading to new types of homeworking, who is doing it, or what the conditions are like, but there seems to be general agreement that it is on the increase. One new group of homeworkers, which has expanded considerably in recent years, is computer programmers. Often, these are mothers who had full-time staff programming jobs before their children were born.

Government cutbacks in spending on nursery provision and other social services mean that more and more women are finding it impossible to go out to work. This, combined with increasingly sophisticated communications technology which links homes with offices, seems bound to lead to more growth in the 'hidden army', as captive homeworkers have been described.

This development gives cause for concern not just to the women directly affected, but to office-worker trade unions, since the use of homeworkers can break down workplace organisation and erode wages and conditions.

What can be done about these problems?

Very little can be done on your own. If you are an office worker who is not in a trade union, you might like to consider joining one.

The checklist of points to cover in negotiations over the introduction of new technology on page 109 shows you what can be done collectively.

If there is already a union for white-collar workers at your workplace, it is fairly simple to join. You simply approach the shop steward or office rep (different unions have different names for the person who represents workers to management).

If there is no union at your workplace, then you will have to decide which one to join, which can turn out to be a little more complicated.

There are two kinds of unions for office workers: those specifically for white-collar workers, such as ASTMS, APEX and NALGO, and 'staff sections' of manual trade unions, of which ACTSS, the white-collar section of the TGWU, and MATSA, the staff section of the GMWU, are examples.

If you work in the office of a manufacturing company, or somewhere else where there is a strongly organised trade union for shop-floor workers, and where you feel it would be possible to form a good relationship with them, it makes sense to join the staff section of their union. In some specialised industries too; there are national agreements which specify the appropriate union for office workers, linked to the shop-floor unions. There are examples of these in transport and mining, and, in the private sector, in the engineering industry where TASS, the white-collar section of the AUEW, is recognised as the union for clerical workers.

In industries that are predominantly white-collar, where shop-floor organisation is weak or non-existent, or where there is a history of bad relations between manual and non-manual workers, a specifically white-collar union is probably more appropriate.

Some of the major white-collar unions are listed below, together with the areas they represent. This list is not exhaustive, and there are other unions that may be right for your workplace.

Public sector unions

CPSA, the Civil and Public Services Association, is the main union for clerical workers in the Civil Service.

NALGO, the National and Local Government Officers' Association, represents white-collar workers in most other public sector offices, such as local authorities, gas and electricity boards, universities and the health service.

NUPE, the National Union of Public Employees is primarily a union for manual workers, but does negotiate for some clerical workers in some parts of the country, for instance in some hospitals.

Other white-collar unions

ACTSS, the Association of Clerical, Technical and Supervisory Staffs, is the white-collar section of the huge Transport and General Workers' Union, which represents workers across a wide range of manufacturing and service industry. White-collar members include some who do not work where there are TGWU shop-floor members, e.g. in advice centres.

APEX, the Association of Professional, Executive, Clerical and Computer Staff, began its existence as a white-collar union. Its members work for private companies of all sizes and for many other institutions. This is the union to which many clerical workers employed by other trade unions belong.

ASTMS, the Association of Scientific, Technical and Managerial Staffs, is well known to many because of the publicity surrounding its General Secretary, Clive Jenkins. Its large membership, which extends into most corners of the economy, is entirely white-collar, with a high proportion of technicians, managers, supervisors and other non-clerical staff.

BIFU, the Banking, Insurance and Finance Union, is, as its name implies, the union for people who work for banks, insurance companies and finance houses. Since it changed its name from NUBE (the National Union of Bank Employees) and widened its scope, it has expanded rapidly.

MATSA, the Managerial, Administrative, Technical and Supervisory Association, is the white-collar section of the General and Municipal Workers' Union, the other giant (along with the Transport and General Workers' Union) among manual trade unions. The GMWU has a very large membership in manufacturing and service industry, which extends into some manual public sector areas. MATSA members tend to work in areas where the GMWU is strongly represented.

SATA, the Supervisory, Administrative and Technical Association, is the staff section of USDAW, the Union of Shop, Distributive and Allied Workers. SATA is the white-collar union for people who work for retailing or distribution firms, such as mail-order firms or the offices of chain-stores, or anywhere else where USDAW represents the shop-floor workers (e.g. in some food companies).

TASS is the Technical and Supervisory Section of the

AUEW, the Amalgamated Union of Engineering Workers. Strictly speaking, its name is AUEW/TASS, and it is a separate union, although it is closely linked with the other three sections of the AUEW. TASS is the appropriate union for clerical workers in the engineering industry, although its membership includes large numbers of drafters, technicians and other non-clerical staff.

You can usually find the address of the nearest office of a trade union in the yellow pages. If you have any trouble finding it, or if you are unsure which union you should approach, contact your local trades council for guidance, or write to the TUC at Congress House, Great Russell Street, London WC1, where there is a special department to deal with such queries.

A note about staff associations

In some workplaces there are organisations called staff associations, which are supposed to represent the workers' interests to management. In a few cases these are in fact genuine trade unions, e.g. the ABS (Association of Broadcasting Staffs) which represents BBC employees, and the GLCSA (the Greater London Council Staff Association) which negotiates for the staff of the Greater London Council. In most cases, however, staff associations are not independent organisations but are effectively controlled or heavily influenced by management. They cannot represent workers as effectively as a trade union can.

3

Factory Work

Manufacturing industry covers a wide range of processes, so working out the prospects for women working in factories is inevitably a fairly complex task. Some products are on the increase, while others are disappearing; some are made by easy-to-automate processes, whereas others still require traditional skills and individual attention. Some factories are owned by thriving, expanding multinational companies, while others are small family businesses perhaps teetering on the edge of bankruptcy. All these factors will affect the job prospects of workers and whether or not new technology is likely to be introduced.

As we have seen, women factory workers, who account for about a quarter of all working women, tend to be heavily concentrated in relatively few industries. Clothing, and food and drink are the biggest employers, followed by electrical engineering, textiles and printing and publishing.

Textiles and clothing are old, established industries, which have been in decline for some time. For centuries, they have provided work for women which has required some specialist knowledge and skill, though never well paid. The most highly paid and prestigious jobs, in both tailoring and textiles, have generally been reserved for men.

In the food and drink industries, too, there has been a distinct demarcation between 'women's' and 'men's' jobs – again with the male jobs being better paid and more highly-skilled. Night-working and carrying heavy weights have often been the pretexts for regarding certain jobs as exclusively for men. Women are to be found in large numbers packing, sorting, labelling, wrapping and carrying out similar operations.

In electrical engineering, there is a similar picture. Here, the largest numbers of women are engaged in the rapid assembly of electrical circuits and appliances, work requiring considerable dexterity and concentration, although it is generally categorised as unskilled or semi-skilled.

Printing and publishing provide yet more examples of the same sort of division of labour. Here, the plum staff jobs are the preserve of skilled workers, gained at the cost of a long apprentice-ship from which women were traditionally barred. Women can be found doing the low-status jobs such as binding, finishing and packing.

None of these industries is entirely escaping the effects of new technology, which is affecting women's jobs in several different ways, some direct and some less so:

● **1. Jobs may be lost when a product becomes obsolete**
Microelectronic components and computerised machinery have provided cheap substitutes for many products that used to be the basis of thriving industries. When the old product becomes redundant, so do the people making it, unless their employer is able to switch very rapidly to manufacturing something else. A classic example of this was the traditional Swiss watch industry, which was devastated within a matter of months by the competition from cheap digital watches made in South East Asia. There are other, less dramatic examples closer to home, mainly in the engineering and electrical engineering industries. Traditional machine tools, for instance, for many years one of the mainstay industries of West Yorkshire, are rapidly being made obsolete by computer-controlled machine tools. Cash registers, mechanical adding machines and typewriters are other examples of products which are being dis-placed by chip-based substitutes.

● **2. Jobs may be lost when a process becomes obsolete**
Microelectronics can be the indirect cause of factory closures, even when the product manufactured there is still a popular one, with customers around the world. When the factory automates the manufacturing process, firms using the new manufacturing methods can produce goods so much more cheaply than firms using

Cheap chips: the secret ingredient

The extreme cheapness of many electronic components is achieved by super-exploitation of the workers who manufacture them, mainly teenage girls in various third world countries. This picture shows women working in an electronics factory in South Korea, where wages are 21 US cents an hour for women (men get 43 cents an hour) and a 7-day, 84-hour week is not unusual, worked in 12-hour shifts. According to one survey, 95 per cent of South Korean electronics workers develop eyestrain, astigmatism or chronic conjunctivitis within their first year of employment, effects which are directly attributable to the strain of working under pressure for long hours through microscopes. To the employers, it does not much matter if the women only last a few months in the job; training is minimal and, with an estimated 40 million unemployed in the underdeveloped countries, there are plenty more where they came from.

Picture by courtesy of Counter Information Services

traditional methods that the traditional-style firms are driven out by the competition. The manufacture of colour televisions provides an example of this. During the 1970s the Japanese TV manufacturers updated the manufacturing process so that by 1977 they had nearly halved the quantity of components in the average 20-inch colour set. These changes, of course, reduced the cost, the complexity, and the production time involved in producing a set, so that it could be sold much more cheaply than an equivalent set made in Britain, where manufacturing methods had remained virtually unchanged since 1970.

One effect of this development was felt in West Yorkshire in 1978 by the people of Bradford and Shipley, when Thorn Electronics announced the closure of two colour television factories, employing 2,200 people, the majority of them women.

The TV industry is only one example of this process, of course. Job loss can come about in a similar way in almost any industry where the production process can become outdated by automation in rival companies.

- **3. Automation can bring about changes in where jobs are**

There is another indirect way in which new technology can cause factory closures and redundancies – by transferring production away from traditional manufacturing areas to other parts of the world. There are several reasons for this. Because parts are now so small and light, it is a quick and simple matter to transport them from one part of the world to another. New technology has also made communications much easier, making it possible for multinational companies to co-ordinate activities in different parts of the world. Finally, by removing the skill requirements from many jobs, new technology has created jobs that can be done by almost anyone, almost anywhere, with minimal training.

These factors combined have brought about what some people have called a 'new international division of labour' – a situation where multinational corporations can stretch their production lines right around the globe, with one part of a product being manufactured here, another there, with sub-assembly in a third country, final assembly in a fourth and finishing in a fifth. The most important factors, in deciding where to produce what, will include the cheapness of labour, convenient access to raw materials,

Hand made by robot

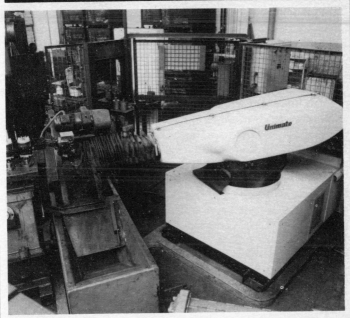

The three-fingered computer-controlled gripper below shows some of the flexibility which robot 'hands' are now capable of, greatly

increasing the range of their applications in manufacturing and assembly. The robot shown above is in operation at the Brook Crompton Parkinson motors factory, presenting an engine part to a stripping machine, after it has been cast. When this machine was introduced, along with other automated equipment, the company's production capacity more than doubled. In other factories, the introduction of robots has not just succeeded in increasing productivity but has also brought about a change to 24-hour production, resulting in more unsocial shift working for the workforce. For operatives working 'downstream' of the robots, a further effect has been a loss of control over the pace of their work, since they now have to work at the speed dictated by the robot.

Picture (above) by courtesy of Unimate and Impact Information Ltd, (below) by courtesy of Computing Europe

the country's tax laws and so on. There are likely to be few incentives for many companies to carry out large-scale mass production of their goods in a relatively high-wage, high-cost country like Britain when there are hundreds of thousands of unemployed people in the third world ready to work for as little as 40p a day doing the same jobs.

Having said this, it is important to remember that not all companies will want to transfer all their production abroad. Many third world countries are regarded as politically unstable, and few companies would want to risk seeing their entire investment lost in a coup. There are other reasons for wanting to keep some production in the advanced industrialised countries – to stay near their final markets, to save transport costs on bulky products, and to protect investments in expensive machinery.

But what has happened in the last few years has made a fundamental change in the bargaining position of British workers in relation to those overseas. It is no longer automatic that jobs will be sited here, and decisions to invest will have to be bargained for. The Thatcher government has provided many incentives such as tax exemption and various forms of grants. In addition, workers will find themselves increasingly under pressure to promise no strikes, low wage claims, high productivity and other concessions in order to keep jobs in their area.

In some cases, the balance of advantage will lie elsewhere. One factory in West Yorkshire provides a sadly typical example of the sort of situation that arises when new technology makes it possible to replace an old product with a newer one, and transfer production to another part of the world. Three years ago, in an electrical engineering factory making components for vehicles, the owners developed a new product, which replaced the complex mass of wiring of the traditional component with a simplified electronic system requiring fewer parts, less time and less skill to assemble. While there is still some demand for the old product, its use is being phased out and it will eventually be entirely replaced by the new one. In the meanwhile, the women who work in the factory have seen their workload shrink until, at the time of writing, they are down to one day a week. Demands that they should be allowed to produce the new product at the old West Yorkshire factory have become impossible to follow through, since their management have now sold a 60 per cent share in the new product to a multinational company, over which they have no control. The chances of continuing work in this factory are almost nil. The most likely scenario is that the new product will be manufactured abroad, and these women find themselves on the dole, their skills useless.

● **4. Automation of production can directly affect jobs**
Needless to say, not all manufacturing companies are prepared to sit back and watch their products or their production processes become obsolete. If they can afford to, many will be looking for ways in which they can introduce new technology themselves to cheapen their production processes or develop new products.

This new technology will itself, in many cases, have profound effects on the jobs in those companies, either by causing redundancies or by changing the levels of skill, knowledge and responsibility required to do the work.

The applications of microprocessors to manufacturing processes are many and varied but there are two broad categories of application which seem likely to have a big impact in the industries in which many women work.

The first of these is *robotics*, a broad term covering a whole range of equipment, where a machine directly replaces a person. As handling and 'sensing' technology becomes more and more sophisticated, there are more and more types of human activity that can be copied and repeated under the control of an electronic 'brain'. One example of robots often shown on television is in car factories, where they are extensively used for spot-welding and for paint-spraying. The machines are 'programmed' by a skilled human operator going through the motions of, for instance, spraying the side panel of a car. The series of actions needed to do this operation is recorded in the machine's memory, and can then be repeated as often as required. Another example is a machine developed by IBM, which consists of a computer-controlled arm with a gripping mechanism equipped with sensors, and a programmeable 'brain' to direct its movements. It can be programmed to assemble a wide range of different products from blenders and toasters to typewriters, staplers and keyboards. For instance it can put together an eight-part typewriter sub-assembly in 45 seconds.

Other handling machines which come into this robot-type category can take over jobs like sorting and packing. Robots can

Such exceedingly good productivity

These Mr Kipling pies pour off the production line at the rate of 40,000 an hour. Fully automatic make-up means that only one worker is left, a solitary woman feeding the foil dispenser.

Picture by courtesy of Spooner Industries Ltd

also be used alongside people, setting the pace and the standard for their work.

The second broad category of microprocessor applications in factories is machinery which carries out *monitoring and control* functions. Many processes in factories demand constant checks and controls on what is happening to various substances. How hot are they? How liquid? What is the proportion of sugar in the mix? Or of dye? How fast is the flow? How quickly is the dough rising? Or the jelly setting? Or the steel hardening? How fine is the yarn? And how thick the paint?

'Sensors' to measure these are being developed very quickly, for applications in many industries. When combined with an electronic 'brain', which can instruct the rest of the machinery to make changes to correct any imbalances, working on the same principle as the thermostat in a domestic water heater, sensors can make it possible to automate control of many manufacturing processes. Automatic process control is not an entirely new phenomenon. In so-called 'continuous process' industries, such as chemical and steel manufacture, where production goes on day and night and stoppages are enormously expensive and wasteful of raw materials, process control has been automatic for some years, using the large, old fashioned mainframe computers. The difference made by the microprocessor is to make this type of operation many times cheaper, and thus open up its possibilities for a number of smaller-scale processes in other industries.

At shop floor level this results in fewer jobs and lower skill requirements, though higher up the hierarchy a few skilled technical jobs may be created.

Of the manufacturing areas we looked at earlier, where women are concentrated, not one remains untouched by at least one of these four effects. In one way or another, new technology seems bound to leave its mark in virtually every area of women's work in factories.

The food and drink industries

These industries provide scope for the introduction of new technology both in process control and monitoring and in robot-type applications.

The preparation of food and drink supplies many instances of prototype systems for automatic process control, from the relatively minor applications like electronic 'dairymaid's thumb' which monitors the rigidity of curd in cheese-making, to major systems which control the entire cooking process for products like biscuits.

Food factories frequently make several different products, and shift from one to another during the course of a day, putting some limits on the degree of automatic process control it is possible to achieve: generally human beings will still be required to shift over from one process to another. In drink manufacture, however, the product tends to be more uniform, and it can often effectively be a 'continuous process' industry, with exactly the same product being made day and night. In such cases the scope for automation is even greater, with the possibility of almost completely automatic production.

The changes which these developments bring about in the relative positions of men and women in the workforce are complex and contradictory. On the one hand, there is a tendency to automate the men's jobs first, presumably because these are more expensive in wage costs. An example of this is the biscuit industry, where the placing of the dough in the ovens, control of oven temperatures, and removing of the cooked biscuits (all jobs traditionally done by men) were automated before the jobs of removing the biscuits from the baking trays, sorting and packing them – the women's jobs. On the other hand, the tendency to move over to 24-hour working results in women workers being replaced by men, who have fewer problems in doing night work.

When it comes to robot-type machines, the jobs currently done by women will be the main targets. These machines are well suited to perform many of the repetitive tasks involved in food production, such as selecting, sorting, grading, arranging, decorating, wrapping, packing and labelling.

To take the confectionery industry as an example, Cadbury's plan to reduce employment in their Birmingham factory from 6,000 to 1,000 over the next ten years with the help of robots, while Rowntree Mackintosh's has begun a feasibility study on the use of 160 Puma robots for sorting and packing chocolates. Another West Yorkshire sweet manufacturer, one of whose factories was represented in the TUCRIC survey on new technology, had introduced robot-style machines into one of its other factories for sweet preparation, with the result that one person was now doing the work previously requiring eight workers.

These developments, and others like them, seem likely to

spread into many parts of the industry during the 1980s with important consequences for many of the women working in food and drink production. However, they will not happen overnight. There are still a number of technical problems to be ironed out before robots can be installed for many processes. For instance, the chocolate-sorting machines referred to earlier have still not been developed with a sufficiently delicate gripping mechanism to avoid marking the chocolates.

Workers in this industry who keep an alert eye out for new developments may well find that they are in a position to anticipate and prepare for new technology before it is introduced.

The clothing industry

Competition from cheaper goods produced abroad has been a problem for Britain's clothing industry for many years, and one of the main reasons for the decline of this industry, now shrunk to a fraction of the size it was at the beginning of the century. Automation has not played a great part in this process. In machining – the job that provides most of the work for women in this industry – the technology has hardly changed for decades. The cheaper price of foreign-made goods can be almost entirely put down to the low wages paid to the workers producing them.

Nevertheless, the clothing industry will not remain untouched by the effects of new technology. In this case, changes seem most likely to come about as a result of robot-style, rather than process-control applications.

Programmeable automatic sewing machines already exist. Once they have been programmed – by recording a skilled machinist going through all the operations necessary to sew a particular garment – they can repeat the operations as many times as required, often at speeds faster than the human operator is capable of.

Although there are reports in Leeds, one of the centres of the clothing trade, of several of the larger manufacturers buying them to try out experimentally, as yet there is little evidence that such machines are being widely adopted in the industry. There seem to be several reasons for this. A skilled machinist has to do a great many things besides operating a sewing machine in order to do her job effectively. For instance she needs to know how to ease together two bits of fabric of slightly different lengths, and to match

Short-circuiting the labour costs

Assembling the components for electrical equipment has been one of the most labour-intensive parts of the manufacturing process in the electrical engineering industry, and one which has, until recently, provided most of the jobs for women in the industry. Now, even this can be automated. The Japanese multinational company, Matsushita, has developed a fully automatic component insertion machine called Panasert, shown here assembling a circuit board for a colour television at their National Panasonic factory in Cardiff.

Below, a woman is taught how to do quality control on the finished product, virtually the only type of job left in the factory.

Picture by courtesy of National Panasonic

patterns. So far, machines are not capable of this degree of sophistication. Another reason is that the garment industry consists predominantly of small firms, usually in old premises with ancient equipment. In recent years, many of them have been hard hit by the recession and foreign competition and do not have the money available to invest in new machinery. Furthermore, many of them are making up garments in small numbers, with considerable variation between styles. In a fashion-dominated industry they cannot predict far ahead what the new orders are going to be like. Thus it is a risky business to invest in machinery which may not be flexible enough to carry out future requirements.

Despite the slow rate of adoption of automatic sewing machines in the industry, it does not do to be too complacent about the unchanging nature of the rag trade.

In the large clothing firms, managements are constantly seeking ways to increase productivity, and several other parts of the manufacturing process are already highly automated. Electronic cutting, for instance, is the norm in some big companies, and there is even a computer cutting service, available on a fee-paying basis to small firms that cannot afford to buy the equipment, or only make occasional use of it.

Cutting was traditionally one of the jobs done by men in the industry. It was relatively well paid and was open only to those who had done an apprenticeship lasting several years. With the introduction of electronic cutting these jobs have disappeared, and are replaced by a smaller number of jobs, done by women working at VDU screens, and requiring only six weeks training. At Hepworths, for instance, all the cutting of made-to-measure suits has been computerised (a central computer is linked directly with terminals in the Hepworths shops, into which individual measurements are fed by shop assistants) and is done by a team of only six women, working in two shifts. When the system was introduced, the company estimated that productivity went up by a staggering 250 per cent.

Hepworths is an example of some new women's jobs created by the new technology. However they are not nearly as skilled as the male jobs they replace; they involve shift working, which puts them out of the reach of many women with domestic responsibilities; and they are very few in number. Six new jobs at Hepworths is a drop in the ocean compared with over seven hundred redundancies created by that same company by the closure of four of its factories in the same year.

The most likely future scenario for the clothing industry is that new technology will halt the trend towards increasing production in the third world, but that this will not result in an overall increase in jobs in Britain, since any new jobs created will be counterbalanced by jobs lost in small firms and in the less efficient parts of the industry. Greater productivity as a result of the new technology will mean that fewer jobs will be necessary to produce the same quantities of goods in the larger, innovating firms.

Electrical engineering

This is the industry most affected by the introduction of new technology. By the end of the 1980s it is likely to have been transformed out of all recognition by changes in all four of the categories discussed earlier in this chapter.

Many factories manufacturing transistor-based appliances, such as radios, TV sets and hi-fi systems, have already been closed, or suffered drastic reductions as a result of competition from more advanced products, based on silicon chips.

Similarly, electrical appliances like washing machines, vacuum cleaners, cookers, irons and fridges are also being manufactured abroad with microprocessor components, indirectly resulting in job losses for British workers making the less sophisticated and more expensive old fashioned products.

When companies do modernise, and introduce microprocessor components into their products, this may still result in fewer jobs, because fewer, smaller components are required, cutting down on the amount of assembly work involved, and on the amount of related work such as storage, sorting, checking and so on. Often, too, there is a change in the skill requirement for the job, as can be seen if you imagine the difference between soldering a large number of wires onto a traditional circuit board, and inserting one tiny chip. This change could bring redundancy to women unable to adapt to the new type of work.

Yet another cause of job loss in this industry could result from automation of the production process itself, e.g. using robots for many of the repetitive assembly-line processes. The workers remaining in the industry would find their work becoming increasingly stressful and monotonous as a result of machine pacing.

Overall, then, it seems likely that jobs will go on being lost in the traditional parts of the electrical engineering industry. However new products are being created in this industry, and new sub-industries growing up as more and more mechanical products are replaced by electronic ones, and new applications of micro-processors developed. Calculators, computer games, word processors and control devices are just a few examples. Surely these new 'sunrise' industries, as they are often called, must provide a secure future and expanding prospects for their workers?

Unfortunately, even here it is impossible to be very reassuring. These industries, mainly in the control of large multinational organisations, or dependent on close associations with them, are among the most mobile the world has yet seen. Research and development (work employing comparatively small numbers, mainly of men) tends to be concentrated in the advanced, developed countries, where these companies have their head offices close to government research agencies, universities and stock exchanges. But the labour-intensive production work may be carried out anywhere in the world.

At the moment, some such work is carried out in parts of Britain where labour is cheap and plentiful and generous development grants and other incentives to invest are laid on by the government, but the bulk of it is done in the Free Trade Zones of the third world. In South East Asia, for instance, silicon chips are manufactured in large quantities in factories employing teenage girls for very low wages, with long hours and often appalling working conditions.

Short of a revolution, only two things could bring about a major shift of this type of work to this country. One would be for wages and conditions here to sink so low that, taking account of other costs such as transportation, it works out cheaper for the companies to manufacture their products here. The other would be for the production process itself to become so highly automated that labour costs become a relatively small proportion of the total cost of manufacturing the products. In the long term, the second of these possibilities may very well come about. The first is highly unlikely to happen.

In the meanwhile, such production jobs as there are in this

Look, no hands!

Automatic carding and blending at the Carrington Viyella Unit 1 mill at Atherton in Lancashire, one of the most advanced textile mills in the world. Fully operational, the mill employs only 90 people to do work previously requiring 430 employees. Job loss was not the only effect on the workforce of the introduction of the new automated equipment: the machinery is run continuously, for 168 hours a week, which means that all workers have to work shifts. In introducing the new shift system, the company 'decided to make use of female labour as far as possible, and the working has been arranged in three shifts of 42 hours week (men) and two shifts of 21 hours/week staffed by women'. (*Textile Month*, June 1978) Clearly the new technology has done nothing to break down the divisions between 'men's' and 'women's' jobs.

Picture: Sefton Photo Library

country for women in electrical engineering are likely to remain insecure, with a constant threat of transfer of the work overseas hanging over the workforce.

The textiles industry

Like the garment industry, textile production in this country has been declining for many years, largely as a result of cheaper production overseas. The industry has seen successive waves of automation over the past three centuries, since inventions like the spinning jenny brought it out of the home and into the first factories. Each wave has tended to remove some of the workers' skill, making it possible for less specialised groups of workers to carry out the work, as well as reducing the numbers of workers required, because of increases in productivity. In part, it was the new technology of the past which made it easy for the British, American and West German textile companies to shift so much of their production away from the traditional textile-producing areas to the cheap labour countries of Asia, North Africa and Latin America. Ironically, it may very well be a consequence of the newest technology that it brings some of this textile production back to the developed countries.

The reason is that microprocessor technology has the potential for automating textiles production to such an extent that hardly any workers at all are required, so the cost of their wages becomes insignificant in comparison with the other production costs.

Microprocessors have a wide variety of applications in the industry. They can be used to monitor and control the spinning, winding, reeling, warping and weaving processes as well as dying and finishing. In fact the all-automatic mill, with one operative controlling up to 120 looms, is now a technical possibility. Nearly all the complex range of skills required of the various groups of textile workers in the past could potentially become redundant.

Having said this, it is important to remember that what is technically possible is not necessarily what will happen. The UK textile industry has not been doing well in recent years, and many companies do not have the money or the inclination to carry out major investment programmes in the near future. However, those who stick to traditional methods are likely to become less and less competitive with their more go-ahead rivals who have re-equipped their mills, so jobs may be put indirectly at risk even when they are not directly threatened by the new technology.

Either way, the prospects for women in the textile industry do not look too rosy, particularly as the new technology also brings with it a trend to more round-the-clock working (because the new machinery is too expensive to leave idle all night) which means that men are slowly replacing women in the few jobs that remain in this shrinking industry.

The printing and publishing industries

The stuff of the printing and publishing industries is information, and one of the effects of new technology has been to break down the barriers between this type of information-handling work and the other kinds, carried out in offices and categorised under a different set of industrial labels in the official statistics. (The transforming effects of microprocessor technology on information-processing work have already been described on pages 18–38 earlier in this book, in the chapter on clerical work.)

If you stop to think about it, the work traditionally done by a typist in an office is remarkably similar to that done by a typesetter in a printing works. The major differences between the two are that the typesetter traditionally takes four or five years longer to learn how to operate a keyboard, and earns four or five times as much per week as the typist. And, of course, 'the print' has been for many years a bastion of male exclusiveness. One of the excuses for keeping women out of highly paid jobs like typesetting in the past has been that part of a typesetter's job has traditionally included carrying around the trays of lead type which can, in old-fashioned newspaper production, be quite heavy (although there is no good technical reason why these trays need be as large as they are).

New technology is rapidly changing this picture. Not only is lead type becoming a thing of the past, but so is the traditional typesetting machine. Computerised production means that text can be fed straight into a computer, using a VDU monitor, to be automatically arranged on the page, in exactly the same way as it is in the production of office documents. The few differences between printing work and office work are rapidly breaking down and the excuses for their being done by different types of workers are becoming fewer and fewer. In existing printing works, it has been rare for there to be a direct substitution of less skilled workers for the traditional workforce. Even redundancies have been unusual. Generally, the price exacted by the unions for the introduction of new machines has been job security for the existing operators, who retain the same wages and job titles as before, even though they no longer need to use many of their skills, and the jobs have become more boring and stressful. However these printing works have lost many of their traditional customers, as, increasingly, large companies and government organisations set up their own internal printing departments, and high-street 'Instant Print' shops flourish, using less skilled workers, often female, paid on scales closer to those of other office workers than of craft printworkers.

Printing employers often seek to exploit this development both to bring down wages in the industry nearer to those in offices, and to worsen the conditions of printing workers. In this they are often unwittingly helped by some sections of the printing unions which, instead of trying to bring women's wages up, so that women can't be brought in as cheap labour, persist in trying to exclude them altogether from the previously all-male areas.

Not all employment in the printing and publishing industries, of course, is in the relatively highly skilled areas of composing and typesetting. Many more employees work as machine operators, printing, stapling, stitching, binding, collating, packing and in a variety of other finishing processes. Women tend to be found, as usual, in the less well paid and lower status jobs, which are also most likely to be defined as unskilled or semi-skilled. It is, of course, also these repetitive jobs that are most likely to be automated.

So far, there is little evidence that this is happening very quickly. Rather, it is more likely to be a long-term trend, starting with the bigger and more progressive companies, and slowly reaching the smaller ones.

Some commentators have predicted that in the long term the printing and publishing industries may very well shrink drastically as electronic forms of communication increasingly take over from the printed word.

In fact, there is little evidence that this will happen. In the past, there seems to have been an opposite tendency – the more experience people have of one form of communication, the more they want of the others too. For instance, contrary to many predictions, the book publishing industry has grown steadily throughout this century, despite the development alongside it of radio, television, glossy colour magazine publishing and, more recently, other media such as microfiches and audio cassettes.

Such commentators too, take little account of the large and growing section of the industry which is not concerned with producing books or newspapers, but with producing things like boxes, wrappers and other forms of packaging.

All in all, it seems unlikely that the industry will shrink in size. That is not to say, however, that the number of jobs will necessarily stay at its present level. New technology may very well reduce it considerably by bringing about increases in productivity, and cutting out some of the traditional processes altogether.

What can be done about these problems?

Very little can be done on your own. If you are a factory worker who is not in a trade union, you might like to consider joining one.

The checklist of points to cover in negotiations over the introduction of new technology on page 109 shows you what can be done collectively. The number of different trade unions involved in manufacturing industry is too great to give an exhaustive list here. Those that follow are just the most important unions in their particular industries.

If you work someplace where there is no union, and are unsure which is the appropriate one to join, contact your local trades council or the Trades Union Congress for advice.

In textiles, clothing and footwear

National Union of Dyers, Bleachers and Textile Workers (NUDBTW), National Union of Tailors and Garment Workers (NUTGW), National Union of Hosiery and Knitwear Workers (NUHKW), National Union of Footwear, Leather and Allied Trades (NUFLAT).

In food and drink

Most workers are in general unions rather than craft unions, mainly the Transport and General Workers' Union (TGWU), the General and Municipal Workers' Union (GMWU) and the Union of Shop, Distributive and Allied Workers (USDAW).

In engineering

The Amalgamated Union of Engineering Workers (AUEW), and the TGWU and GMWU.

In printing and publishing

The Society of Graphical and Allied Trades (SOGAT), the National Graphical Association (NGA), the Society of Lithographic Artists, Designers, Engravers and Process Workers (SLADE), the National Society of Operative Printers, Graphical and Media Personnel (NATSOPA), and the National Union of Journalists (NUJ). *Note*: At the time of writing, some of the print unions are discussing possible amalgamation.

4

Other People's Housework

As we have seen, a very large number of women – nearly a quarter of all working women – work in occupations that can be described as 'housework' – cleaning, cooking, washing and waiting on people.

These are not the sorts of jobs that make headline news when they disappear. In fact, usually they are so little noticed that many people hardly realise they exist. Yet every time a factory, an office or a large shop closes, there will be redundancy notices for several cleaners, canteen assistants, lavatory attendants and other ancillary staff. It will be rare, however, for them to get a fraction of the redundancy compensation that other workers in that workplace will get.

There are several reasons for this: in any given workplace these ancillary workers will be in a minority and therefore not in a good position to bring their demands to the forefront; they are also likely to be part-time workers, ineligible for many of the rights full-timers take for granted. Often they will not even be in a trade union, since specialist craft unions will not take them into membership in many industries.

A further problem, and one which is unfortunately on the increase, is that many cleaning and catering workers do not even work directly for the boss, but are employed by a specialist contract agency. This adds to their insecurity, and places more barriers in the way of organisation.

There is another, more insidious, reason for the extremely low status and low pay of this type of worker, directly connected with the type of work they do. It isn't that the work is particularly

Cleaning:
the endless round

For many women, like this London nursery cleaner, the only available paid work is almost indistinguishable from the unpaid work they do every day at home. Like one's own housework, other people's cleaning and catering carries little job satisfaction; each day, what you have done is undone again, to be endlessly repeated. In addition, it is exceptionally poorly paid, with bad working conditions and little job security.

Picture: Margaret Murray

easy, or light, or that it demands no skill or knowledge to do properly; it's just that it's – well, it's housework. And housework is something all women are supposed to be able to do effortlessly. After all, they've been learning to do it since they were toddlers. Furthermore, it's work you normally don't expect payment for. Who ever heard of a housewife being paid to do the washing up, or to scrub the kitchen floor?

So a woman who does this sort of work for a living is bound to find on the one hand that her work is grossly undervalued – she's often made to feel that she's pretty privileged to be paid at all – and on the other hand, that her work has no scarcity value.

She can always be replaced by any other woman desperate enough for a few pounds to accept any work, and has little bargaining power compared with other workers.

Generally speaking, as we have seen, these workers tend to be in a minority in fairly large organisations, which makes it particularly difficult for them to organise effectively. There are, however, some industries where they can be found working together in large numbers. These are the hotel and catering industries, and other areas where catering is carried out on a large scale, such as hospitals, school meals services and airports.

In some of these areas, progress has been made in recent years in introducing trade union organisation and bringing about improvements in wages and conditions. Unfortunately, however, these same areas are most vulnerable to the inroads of new technology.

Cooking, preparing and serving food are all activities that are becoming increasingly automated. School meal services, social services and the health service have been turning to systems where all cooking is done in one central computer-controlled department and then sent out to be heated in microwave ovens on the spot where it is to be finally consumed, thus eliminating jobs in individual schools, day centres, meals-on-wheels depots and so on.

In hotel and restaurant chains, computer control of timing and temperature have paved the way for the rapid expansion of the 'fast food' chains, while sophisticated dispensing machines have enabled some canteens and waiting-room buffets to go entirely over to self-service, substituting coin-in-the-slot machines for cooks and waiters, and throwaway cups and plates for dishwashers.

In hotels, computerisation does not just stop at the kitchen

The automatic canteen lady

Even housework-type jobs can be automated. At Pedigree Petfoods, in Peterborough, main-course dinners are now provided by this vending machine, known as 'The New Grand Gourmet Refrigerated Merchandiser'. Having selected and helped themselves to their meal, staff then heat it up in a micro-wave oven, thus cutting out much of the work previously done by catering staff.

Technology like this has widespread applications in schools, hospitals, transport cafeterias, commercial catering and anywhere else food is prepared and served on a large scale.

Picture by courtesy of National Vendors

controls but can be extended to cover virtually every function from taking and dispensing drinks orders to keeping records of who has consumed what.

Automation in catering does directly destroy some jobs, and de-skills many others. However its greatest effects are probably felt elsewhere, in the small traditional cafes and restaurants that are pushed out of business by competition from the fast food giants.

In other industries, 'housework' workers suffer indirectly wherever new technology destroys the jobs of other workers. Fewer people working in factories and offices means fewer to cook for and fewer to clean up after. And all this spells fewer jobs for women who have only their housework skills to offer on the labour market.

What you can do about these problems

Turn to page 109 for a checklist of trade union demands, if you are already in a union. If you are employed in a 'housework' job, and are interested in joining a trade union to protect your interests, the most appropriate one will depend on where you work:

In the hotel and catering industry

The General and Municipal Workers' Union (GMWU) and Transport and General Workers' Union (TGWU) have special hotel and catering sections, one of which is probably the most appropriate one for you. There are a few exceptions to this. If your employer is British Rail, then the National Union of Railwaymen (NUR) is the appropriate union. In some other establishments, where other unions are already strongly organised, these may be the best ones to join. For example, in some leisure complexes which include cinemas, NATTKE, the National Union of Theatre, Television and Kine Employees, represents the workers, and in some transport catering establishments, the Transport and General Workers' Union (TGWU) is recognised for all staff.

In the public sector

NUPE, the National Union of Public Employees, represents most cleaning and catering workers in schools, hospitals and other public services. However in some parts of the country these workers, particularly cleaners, are members of the GMWU, the General and Municipal Workers' Union.

In other areas

In many cases, where 'housework' workers are in a minority, the best organisation to further their interests is the trade union that represents the majority of workers at that workplace. For instance, in mining, the National Union of Mineworkers (NUM) represents canteen workers as well as miners. However there are some specialist unions that will not recruit 'general' workers into membership, and other workplaces where, even though the union is theoretically prepared to take in cleaning and other 'service' workers, these workers don't believe they will get a fair hearing in that union or that it is in their interest to join. In such cases, the most appropriate course of action would be to approach one of the 'general' unions – the GMWU or the TGWU – for help.

5

The 'Caring Professions'

The fourth major area where women work is in the so-called 'caring professions' – teaching, nursing, nursery nursing, social work and allied jobs. Like many other jobs done by women, these can be seen as extensions of housework. Women are supposed to be naturally suited to them because 'caring' is automatically expected of every woman in her role as daughter, wife or mother.

There are several characteristics of these jobs, though, that make them different from the other 'housework' jobs.

Firstly, they are 'professional'. That is, they require particular skills and training, which often take several years to acquire. This has made it possible for these workers to develop some degree of job security over the years, compared with other women workers, since they can only be replaced by others with similar skills and training. It has also facilitated the development of 'career structures', offering some chance of progress and improvement within the professions, with relatively well paid and high status jobs at the tops of the career ladders. Needless to say, where you find well paid and high status jobs you are also likely to find men, and one of the characteristics of these professions is that, although in some cases, like nursing and infants teaching, they started off as entirely female, they now have increasing numbers of men in the authority-exercising positions at the top – head teachers, hospital administrators, social work managers etc.

Secondly the 'caring professionals' are working directly with people with whom they have a special involvement, very different from that in most other jobs. Usually they feel directly responsible for the welfare of others – often people who cannot effectively look after their own interests because they are young, sick, old or socially inadequate. This makes it very difficult for these workers to take the sort of action which other groups might take to defend their interests. To go on strike, for instance, becomes a much harder decision if it might involve hurting helpless people whom you really care about. These workers are usually very committed to their work, and often put in extra hours of voluntary effort or take on additional duties without dreaming of claiming overtime pay if it seems to be in the interests of the people they serve.

A third characteristic of this group of workers is that most, though not all of them, work for the state. The work they do is not so much dictated by the laws of supply and demand (or 'market forces' as the Tories like to call them) but by government policy, at national or local level. And productivity cannot be measured in output, but must remain a subjective judgement, based on professional standards. The future of these jobs, therefore, is likely to be determined by political decisions – decisions about the levels of public spending that will be allowed, about the type of services that should be provided and so on. An unemployed nursery nurse, for instance, is likely to get a job not because there are large numbers of women in the area desperate for daycare for their under-fives, but because the local council has taken a decision to spend money on a nursery.

Political factors, therefore, will decide how many jobs there are going to be in these services in the future, and new technology is unlikely to have any direct effect on the *number* of jobs. However it does seem likely that, at least in some of the 'caring' services, new technology will bring about some changes in the *types* of work carried out, and even in the proportions of men and wmen doing these jobs.

Education, social services and the health service are all areas in which professional workers are supposed to collect a good deal of information about their pupils, clients or patients. The actual handling of this information – the work that can be done directly by machines – is done by clerical and administrative workers, whose jobs are likely to suffer, as we saw earlier (pages 18–38). However it is the professionals who generally *provide the information in the first place*. It is they who mark the papers and write the school reports, who take and record medical readings and who write court reports

Caring:

the self-imposed restraint

As mothers, daughters or wives, most women spend much of their time caring for others, and this expertise as carers is exploited in many of the jobs they do. The very fact that they do care about their charges can make it extremely difficult for them to take militant action to improve or defend their jobs. This Islington childminder, for instance, could never just down tools as a factory worker could, and abandon the children in her care.

Picture: Margaret Murray

and keep other records on those in the care of social services departments.

The computerisation of this information has several effects. Although it may speed up some routine work, it can considerably reduce the job satisfaction of many professional jobs by simplifying what have in the past been complex matters of professional judgement, making them crude yes/no pieces of information that can be handled electronically. When a social worker is asked for a report on a client charged with a criminal offence, for instance, or a teacher asked for an employer's reference on an ex-pupil, it is usual for a personal opinion to be given, in the professional's own words, carefully chosen for that particular individual. With computerised records, it is increasingly likely that such data will simply be pulled out of the machine's memory, perhaps even without the knowledge of the person who made the original report, with information that has become distorted through being made to fit the pre-existing categories on an unsuitable standard form. It may be, perhaps, that a child has been through a phase of maladjustment at school due to problems at home. On a form that simply asked for the answer yes or no to a question about adjustment to school, this child might end up being classified in the same category as a seriously disturbed fellow-pupil with chronic problems.

Developments like these could seriously damage the quality and variety of work for many professionals, making their jobs more routine and removing responsibility. The same processes could also lower the quality of the service for the clients, leading to inhumane and inappropriate treatment, and to errors that could be avoided by the exercise of an individual worker's common sense and professional judgement.

Perhaps even more worrying, from the point of view of the users of the services, is the loss of confidentiality, which comes about as soon as a lot of information is held in the memory of a computer to which numbers of people have access. It is possible to imagine a time, not too far in the future, when all the information about an individual, from birth onwards, is held in a network of inter-communicating machines. This could include embarrassing medical details, criminal records, records of behavioural problems at school, and information that is not even factually correct but that is, perhaps, the personal opinion of an individual teacher, psychologist or social worker. No one is infallible, and the caring professions, like all others, include their fair share of religious bigots, racists, sexists, anti-homosexuals etc. whose opinions could be very damaging. To be concerned about the effects of the prejudices of individual professionals is not to contradict what was said earlier about the importance of individual professional judgement. When professional opinions were given only in individual face-to-face meetings and one-off reports, then any idiosyncracies of the particular professional would be apparent to the others involved with that particular case, and there was a chance for others to correct the bias. The dangers of computerised information banks lie largely in their anonymity and in the spurious 'objectivity' of computer information when the person who put it in is invisible. Other problems can arise when computer records are used for purposes quite different from those for which the information was originally intended.

As people with a commitment to the welfare of the community and to their own professional standards, many caring professionals find cause for serious concern in these developments. But these are not the only ways in which new technology can affect their working lives.

In the health service

New technology brings with it an acceleration of the trend towards increasingly 'scientific' nursing, i.e. nursing which is more and more dependent on machinery and less and less on human care. In intensive care units, and some obstetric units, nurses and midwives are already finding that their jobs include more machine-minding than interaction with patients – presumably not what they joined the profession to do. Apart from lowering the job satisfaction for many nurses, this trend has two other consequences that are important for women: By making medicine more capital-intensive, it accelerates the trend towards big centralised hospitals at the expense of smaller local units, which were conveniently near homes and enabled many women to arrange part-time working. It has also been blamed for an increasing trend towards employing men rather than women in the higher reaches of the profession, since they are more likely to have the scientific or technical training which is becoming increasingly important in a machine-dominated health service. This seriously reduces the promotion prospects for women in a profession which, traditionally, provides very few rewards for

Innocent spies

The stuff of many of the 'caring' jobs is information – information about many of the most private and intimate parts of our lives: our medical records, social problems, learning difficulties, brushes with the law or even – as is the case in VD clinics – our sexual encounters. As more and more of this information is computerised, the hard-pressed women who carry out so much of this sort of work in the welfare state may find themselves unwittingly becoming part of a massive intelligence-gathering network which seriously threatens the civil liberties of the population.

Picture by courtesy of Computing Europe

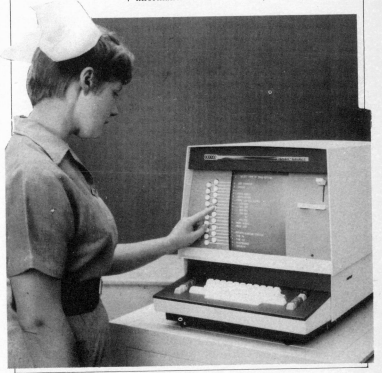

those at the bottom. In the past, many women stuck it out for their first few years only because they could rely on eventually becoming a sister, or reaching some other position of relative importance, if they worked well enough. In the future, this channel may be closed, for all but a small minority.

In education

Although teaching machines have been around for some time, it is highly unlikely that technology will directly take over the jobs of teachers, at least teachers of those below school-leaving age, if only because such a large part of the schoolteacher's role in our society is that of social control – keeping the kids off the streets, in some semblance of order, and instilling habits of punctuality, obedience, politeness and respect for authority – tasks no machine yet designed can carry out effectively.

It is likely, however, that the increasingly electronic nature of so many different aspects of society will bring about changes in many teachers' work. Apart from being expected to prepare their pupils for a working life that will probably include long stretches of unemployment and require different social attitudes, they will also have to teach them much more in the way of basic computer-related skills. Since men are much more likely to be qualified in these scientific and technical areas, as this table shows, it will give them a competitive edge over women teachers when applying for jobs or

Students on Engineering and technology courses, 1977		
	Women	Men
Full-time polytechnic students enrolled on advanced courses	275	5,102
Full-time university students on undergraduate courses	1,657	32,086
Full-time university students on postgraduate courses	457	6,396

Source: *Statistics of Education*, Department of Education and Science

Men are more likely than women to be qualified to teach computing and the other technical subjects for which there is likely to be an increased demand.

New clients for the

social worker

The combination of rising unemployment and declining public spending spells crisis for the inner cities, as was shown during the riots of the summer of 1981 when this picture was taken in Toxteth, Liverpool. As violence increases and mental illness and suicide rates climb, it is social workers, community workers and probation officers who have to pick up many of the pieces. At a time when their numbers are being cut, this can add almost intolerable strain to an already difficult job.

Picture: Laurie Sparham Network

for promotion. At a time when the numbers of teaching jobs are shrinking anyway because of public spending cuts and a falling school population, this will reduce still further the job prospects for women teachers.

In social services

Here the effects of new technology are likely to be more diffuse. It is just one factor among many in the increasingly crisis-ridden situation with which social workers are having to cope, at a time when their resources are actually being cut back because of government policies. Rising unemployment, deteriorating housing and declining services and other by-products of the recession and the cuts all contribute to the rising crime rate, increase in domestic violence and other social problems with which social workers have to deal.

Present government policies seem designed to transfer increasing amounts of this work onto volunteers, and turn social workers into more specialised 'experts'. Nevertheless, the workload will certainly continue to increase as services are stretched to breaking point, and social workers will inevitably find their work becoming more and more stressful as they continue to be blamed whenever a child is battered to death or another wretched casualty of the system comes to grief.

Internationally, some commentators have predicted that social work will grow as a profession as the industrialised countries increasingly become 'leisure' (or high unemployment) societies as a result of automation, but this seems unlikely to happen in Britain unless there is a complete about-turn in government social policy which has, until now, been committed to reducing expenditure in this area.

In summary, it seems as though the caring professions will not be affected by new technology in as direct and immediate a way as most of the other types of work women do. However women in these jobs are likely to find themselves having to contend with many of the gravest of the indirect effects of automation. Not only will they have to pick up the pieces of the broken lives that technology has shattered, but they will also increasingly find themselves forced into the role of information police for a technological state.

What you can do about these problems

In addition to some of the job-protecting strategies suggested in the checklist on page 109, trade unions in the caring professions could play a part in developing demands to protect professional ethics and confidentiality of information in the public services. It might be possible, for instance, for groups of teachers, social workers or medical personnel to consult with groups representing users of the services and draw up guidelines or codes of conduct to control the type of information recorded about individuals, and access to those records.

These demands could be strengthened by linking up with civil liberties campaigns to develop public awareness of the problems and broaden your campaigns.

Some useful organisations to contact are:

National Council for Civil Liberties, 186 Kings Cross Road, London WC1 (01-278 4575)

State Research, 9 Poland Street, London W1 (01-734 5931)

The main trade unions for the caring professions are as follows: (Please note: This list is not exhaustive, and does not cover all the smaller groups such as probation officers or speech therapists who are included in this category of work.)

National Union of Teachers (NUT) covers teachers and nursery teachers in England and Wales.

The Education Institute of Scotland (EIS) is the Scottish equivalent of the NUT.

NALGO, the **National and Local Government Officers' Association**, represents social workers, nursery nurses and some other professionals employed in the health service and in local government.

NUPE represents some lower grades in social services, such as care assistants.

In the health service, the trade union structure is more complex; the main unions for nurses are **COHSE**, the **Confederation of Health Service Employees; NUPE**, the **National Union of Public Employees**; and the **RCN**, the **Royal College of Nurses**, a professional association rather than a trade union.

Checking out

the back-room jobs

This checkout operator is working with an electronic point-of-sale terminal. Instead of keying each purchase into a till, she passes the bar code (below)

printed on each item over a little slot on the counter, through which a laser beam passes and 'reads' the information on the code. When items have to be weighed, that too is done electronically using the digital scales

in the right foreground of the picture. The recorded information is not just used to tot up the customer's bill, but also for stock-keeping and other accounting purposes, making many of the traditional 'behind the scenes' shop jobs redundant. This type of system, comparatively new in this country, has been in use for some years in Scandinavia, the United States and Canada, where the customer's receipt (below) comes from, showing the type of information recorded on these machines. Connected up with details of the customer's bank or credit card

account, this information could provide yet another source of intelligence about the individual to surveillance agencies.

Picture by courtesy of Key Markets

```
ZEHRS  WATERLOO SQUARE

HORSERADISH MST          .51
3LB DEL APPLES         1.59
CARROTS 2LB BAG          .50
GROCERY                 1.12
BANANAS                  .83
2.50 LB AT .33/LB
GOURMET COFFEE         2.44
.51 LB AT 4.79/LB
SCOTCH BRITE             .72
CANADA VINEGAR           .71
2 KG SUGAR             1.39
TOMATOES - USA         1.25
HOMO MILK JUG          1.89
DEPOSIT                  .50
COMET CLEANSER           .63

TAX                 $    .05
TOTAL               $  14.23

CASH TND            $  20.00
CHANGE DUE          $   5.77

CHECK-OUT 5   TS-57
11/2/81 16.34        18983
THANK YOU, COME AGAIN
```

6

Sales and Distribution Work

Shop work has been radically transformed over the last few decades, with small local shops, each with its owner, being replaced by large supermarkets and chain stores, and with individual selecting, demonstrating, cutting and packing of goods by a skilled assistant being ousted by mass-production and pre-packaging of branded goods. The delivery boy on a bicycle, the milliner, the pastrycook, the cobbler, the fishmonger and the many other shop workers our grandmothers encountered on a typical trip down the high street are now all rarities. They have been replaced by the stores clerk, the shelf-filler and the cashier, with a hierarchy of managers above them. Many small shops still exist, of course, but their numbers are shrinking fast, with a record level of bankruptcies in the last few years. Where they survive, if they are not family businesses, they increasingly rely on unskilled, often part-time, staff.

In recent years, a massive growth in part-time working in retailing has made up for and obscured how many full-time jobs have been lost. Nationally, in the five years between 1971 and 1976, 171,000 part-time jobs were created in retailing while 106,000 full-time posts disappeared. While the big chains of stores have been enlarging their empires at the expense of the small shopkeepers, other types of selling, notably mail-order, have also been expanding. Both require a good deal of administrative work, and until recently have been employing increasing numbers of clerical workers. These large organisations also require central warehouses in which they employ assemblers, packers, handlers etc. These two groups of

The small shop: the
first victim

It is likely that most of the job loss in shops caused by the new technology will occur, not in the big supermarkets and department stores where the new machines are introduced, but in the small shops, which cannot afford to compete with them and are driven into closure as a result. With survival already precarious for many of them, the extra competitive edge given to the big shops by automation can often be enough to tip the small ones over into bankruptcy. The traditional London corner shop in this picture has already disappeared.

Picture: Margaret Murray

workers make up a significant proportion of the totals usually listed in the statistics as 'retail and distribution workers'.

Selling and distribution involves keeping track of the details and whereabouts of vast numbers of different items, in varying quantities – a huge amount of record-keeping, routine communication and arithmetic, all of which are natural candidates for information technology. In addition, the movement of goods into and out of warehouses is a relatively simple process to automate, using conveyer belts with microprocessor control and robot-type handling devices with microchip 'brains'.

Thus there are a variety of ways in which new technology can affect selling and distribution work.

POS—the electronic cash register

POS is an abbreviation of Electronic Point of Sale Data Capture Equipment, a computerised replacement for the cash register or till at the supermarket checkout or the pay desk in the department store. However it does much more than just record the amount and store the cash. It is designed to 'capture' information at the same time. This information varies depending on how the machine is programmed and what information is keyed into it, but it could include any or all of the following: what item has been sold, together with any details such as stock number, discount, colour, model number etc.; details of the customer (if an item has been paid for by cheque or credit card); details of the checkout operator and how fast she is working, how often she makes a mistake and when she takes a break. Information about the product can either be keyed in by the operator or automatically 'read' by a laser scanning device from a coded magnetic strip on the product label, known as a bar code. Most items on sale in British shops already have these bar codes attached.

Once the information has been 'captured' it can be transmitted automatically to other computers. To the warehouse, for instance, to order more stock of something that has nearly sold out; to the accounts department, to communicate how much money has been taken; to the personnel department, to update the records on individual workers' productivity. In the United States, POS are already used to communicate directly with customers' banks, and deduct money automatically from their accounts to pay for the

The automatic warehouse

This picture shows one of the first automatic warehouses in this country, at Penguin Books in Harmondsworth, Middlesex. The uniform nature of the product – standard-size paperback books – and the large scale of the operation made this a suitable application for the use of a traditional mainframe computer for automatic control of the conveyor system, and for stock-keeping and order processing. Now, the use of cheap, flexible microprocessor-based computing has made it possible to extend these sorts of systems to many other types of warehouses, including those which stock a wide range of differently sized and shaped goods.

Picture by courtesy of Computing Europe

goods they have purchased. This is known as electronic funds transfer, or EFT. All of these forms of communication, of course, replace traditional paper-based ones, which employed a large number of workers – e.g. stock-keepers, order clerks, invoice clerks and book-keepers. Many of their jobs could now be at risk.

The use of bar-coding on products can also get rid of the job of price-labelling the goods on supermarket shelves. Since each article is already coded with its price, all that is needed is a single price-tag on the shelf – much less labour-intensive than pricing each item separately.

The effects of POS technology on shop workers are twofold. On the one hand it can change the nature of the checkout operators' job and make it subject to continuous monitoring. On the other, it can cause job losses among other sales workers and clerical support workers in retailing companies. Most commentators believe that it will not lead to massive redundancies in the firms that introduce the new technology, because the increase in productivity they will get will enable them to keep down the prices of their goods, thus making them more competitive and giving them a bigger slice of the market. It will be in the smaller firms, which do not put in new technology, that the greatest job losses will be felt, as more and more of them are forced into closure by this competition.

Since employees in the small firms are least likely to be in trade unions and to be covered by redundancy agreements, this will fall especially hard on these workers.

Other types of automation

The effects of automation on warehouse workers will also be great. In the TUCRIC survey, there was an example of a warehouse that was about to go fully automatic, with machine control of belt operation, stock movement, stock control, assembly and packing. Stock-keepers' jobs were to be eliminated as a result of the computerisation of stock ordering and control, and 14 women's jobs were due to disappear as a result of the computer-controlled belt movement. Assembly workers were to be retrained and regraded and extra stress was predicted for those who remained.

All in all, sales and distribution workers seem likely to suffer major changes as a result of the introduction of new technology. Clive Jenkins and Barry Sherman of ASTMS estimate that as many as 40 per cent of jobs in this sector will disappear in the long term. While others are reluctant to specify an exact number, there seems to be agreement that there will be some decline in employment. For the workers left behind, it looks as though most jobs will become less skilled and more stressful, unless steps are taken to prevent this happening.

What can be done about these problems?

If you are a sales or distribution worker who is not in a trade union, you may wish to consider joining one. The appropriate trade union for most workers in this sector is **USDAW** the **Union of Shop, Distributive and Allied Workers.**

The checklist of points to cover in negotiations over the introduction of new technology on page 109 shows you what can be done collectively.

7

What New Jobs Does New Technology Create?

Up to now, we have mainly discussed the effects of new technology in destroying jobs – and a fairly dismal catalogue it is. But it must also create some new jobs. What hope will these provide for women seeking work in the future?

In the TUCRIC survey of workplaces in West Yorkshire, there were ten cases (out of 17 answering this question) where new jobs had been created after the introduction of new technology. Almost without exception, these were jobs directly connected with computer programming and systems analysis. The exceptions were one case where four electricians' jobs had been created (presumably for maintenance and installation work on the new machines); one case where extra stores clerical assistants were taken on (probably because of expanding business rather than a direct result of the new technology) and one case where word processor operators had been taken on (the usual practice seems to be to retrain typists as operators rather than taking on new staff for this work). In no case were these new jobs created in large numbers (although they were often in places where large numbers of jobs had disappeared as a result of the new technology).

If these findings are typical, it seems likely that substantial numbers of new jobs for skilled computer staff are being created around the country at present, a supposition borne out by the fact that experts in the electronics industry and in the government are seriously worried about a 'skills shortage' in computer-related occupations.

So, could these vacancies become available for some of the

hundreds of women being displaced from other occupations by the changing technology? The answer seems to be that this is highly unlikely to happen. Although women are being given some basic training to enable them to fill the low-grade office jobs connected with new technology, such as word processor operating, only a tiny minority get the more specialised training which would enable them to fill these higher-level technical jobs, as the table shows.

Men and women completing state provided training in 1978/79

Type of course	Women		Men	
	Number	%	Number	%
Clerical and Commercial	22,156	91	2,162	9
Management	1,208	19	5,110	81
Science and Technology	94	9	900	91
Craft Skills	314	1	28,228	99
All courses	28,332	40	42,033	60

Source: Hansard, 1979, quoted in *Microelectronics* CSE Books 1980

The most likely scenario seems to be that these new jobs will be filled overwhelmingly by men, with women concentrated in the low-paid ghetto at the bottom of the office hierarchy.

Another category of new employment created by the new technology is in the manufacture of new products, in the 'sunrise' industries, generally parts of the electrical engineering industry, some of which were touched on in the last chapter. This work falls into two main types. There are highly paid jobs demanding a lot of skill and knowledge in research and development. Because of the educational factors already outlined, these will almost certainly be predominantly male. In this country, they are also likely to be geographically concentrated in London and the Home Counties where many companies have their head offices, or near major universities or other areas with resources for carrying out research in electronics, for instance the area around Cambridge.

Then there are the jobs involved in mass-producing the finished product, which tend to be unskilled and poorly paid. As we have seen, although these may provide a few jobs in depressed regions in the short term, they are unlikely to provide large numbers of secure jobs in this country while the cheap labour of the underdeveloped countries continues to provide such an attractive alternative to the employers.

In some industries, other types of jobs are opened up to women by the introduction of new technology. Examples we have touched on include the operation of electronic cutting machines in the tailoring industry, and new printing jobs in large companies and government departments. Strictly speaking these are not 'new' jobs, even if they are new for women, because they replace jobs previously done by men. As technology increasingly supplies electronic substitutes for craft skills and physical strength, it is likely that other types of work will be opened up to women in this way, for instance in the engineering industry. It should be remembered, however, that the numbers of jobs 'created' are usually far fewer than those lost. Women in industries where such changes are taking place should also be alert to the dangers of a divide-and-rule strategy by the employers. It is no advance for women to be admitted to previously all-male bastions of employment if the price for this is rock-bottom wages and the destruction of trade union organisation. Similarly, men should be reminded that the way to protect the wages and conditions they have won in the past is not by excluding women from their organisations, but by including them, and insisting that they are paid equally.

In the longer term, it is possible that other types of new jobs will be created, in new services which grow up to supply the new needs of an increasingly electronic society. We can speculate, for instance, about the prospects for floppy-disc libraries, word processor servicing agencies, eye-testing services for VDU operators, electronic baby-alarm agencies which baby-sit at long distance and many other more, or less, alarming possibilities.

It is also probable that there will be new manufacturing industries, making as yet undreamt of products. How and when they come to pass will depend on major economic and political factors outside the scope of this book. At present, all that can be said with any degree of certainty is that, while the current recession lasts, and the policies of the major Western governments remain unchanged, there seems little likelihood of new technology-created jobs making more than the tiniest dent in the unemployment figures.

8

How New Technology Affects All Women

So far, we have looked at some of the ways in which new technology affects specific jobs done by women. There are, however, other ways in which the effects of its introduction are more far-reaching, bringing about changes which can affect all women, regardless of the particular work they happen to do.

Firstly, rising unemployment affects women much more seriously than men. Between 1975 and 1980, female unemployment rose at three times the rate of male unemployment, and in 1981 it was estimated that nine women were still losing their jobs for every five men who became unemployed. These figures, based on official statistics, may very well be a severe *under*estimation of the real extent of women's unemployment, since there are several reasons why some women don't sign on, and thus never appear in the official figures—for instance, because they are only paying the 'married woman's stamp', and thus aren't eligible for unemployment benefit, or because they are part-time workers, for whom signing on is discretionary.

One of the reasons for this higher rate of job loss among women is that women have a higher turnover rate than men anyway. Because they are expected to put their families first, they tend to leave jobs sooner and more frequently. So that when an employer has an across-the-board 'natural wastage' policy, women's jobs are the most likely to disappear. In many large organisations such as the offices of major companies, local authorities or colleges, for instance, a target of say 5 per cent staff reduction by natural wastage will be met almost exclusively by clerical workers leaving –

meaning that the remaining clerical workers are carrying the burden of the extra workload.

Natural wastage is the way that most jobs disappear at the moment in this country. In the TUCRIC survey, out of 29 workplace representatives answering this question, no less than 24 said that their employers had a policy of natural wastage.

While this method of getting rid of jobs might seem preferable to redundancy in that nobody is actually deprived of a job against their will, it does nevertheless have serious effects. The woman who leaves may very well be doing so not because she wants to leave the workforce for good, but because she can't find anyone to look after her children over the school holidays, or because her husband has got a job in another part of the country, or because a parent is dying and needs full-time nursing. The disappearance of the job also means one fewer vacancy for all those others who are looking for work.

Natural wastage is not the only way that women lose their jobs faster than men. In the TUCRIC survey, people were asked what principles operated in their workplaces when redundancies occurred. Several workplaces operated a combination of principles.

Natural wastage	**24**
Voluntary redundancy	**10**
Early retirement	**7**
'Last in, first out'	**6**
Part-timers first	**6**

The last three probably come into effect when there are not enough volunteers to achieve the desired job loss by the first two methods.

Of these different methods for selecting people for redundancy, two are clearly discriminatory against women – 'last in, first out', because women are likely to have been in their jobs for shorter periods than men because of domestic responsibilities causing interruptions in their working lives; and 'part-timers first' since the vast majority of part-time workers are women, again because of their family commitments.

Natural wastage and getting rid of part-timers and recent recruits are also ways in which the employer can make redundancies extremely cheaply. People who leave of their own accord do not have to be paid any redundancy money at all, while part-timers and recent recruits are likely to be entitled to amounts of money which are either small or non-existent, as there is no legal entitlement to redundancy pay for full-timers who have worked in a particular job for less than two years, or part-timers who have been employed for less than five. Not only does this mean hardship for the individuals who have been made redundant, but it also spells bad news for the union at the workplace, which has 'sold' these jobs for next-to-nothing, a precedent that will undermine its bargaining power for the future.

The discrimination against women, which takes place when jobs are disappearing, is bolstered by male attitudes that regard women as dispensable. It is all too common, still, to hear women accused of taking the bread from the mouths of working men, even when these same women are the sole breadwinners for their families, and for it to be assumed that the only unemployed people the government should be concerned about are men.

The same discriminatory attitudes can also mean that, once unemployed, women have a harder time finding another job, if they are applying for the types of work in which they are in direct competition with men.

Apart from the general effects on unemployment, there are other, more subtle ways in which new technology can make life more difficult for women. These are connected with the increasing automation of service employment, some of which has been described in earlier sections of this book.

A very high proportion of most traditional 'service' work has been concerned with directly 'serving' members of the public, whether this has been in shops or banks, in petrol stations or restaurants, in hospitals or government offices. When new technology is brought in to increase the productivity or efficiency of these 'service' workers, their jobs change: the work is speeded up, a higher proportion of it devoted to tending machines, and a lower proportion devoted to the direct face-to-face 'serving' of the general public. Instead of dealing directly with a worker, customers find themselves faced with a vending machine, a cash-dispenser, a self-service counter, a ticket machine or a queue at a desk where one super-efficient worker at a computer terminal or a point of sale terminal is achieving 'maximum customer throughput' with a minimal amount of labour.

The 'service' that used to take up so much of the working time of service sector workers has not, in fact, magically disappeared as a result of these new machines, although many of the jobs may have gone. What has happened is that service by a paid worker has been replaced by *self*-service by the customer. A task that used to be part of somebody's paid job has been transformed into an unpaid part of somebody else's consumption. Employers like to talk of the labour they have 'saved' as a result of introducing new technology. A more accurate term would be 'off-loaded'.

This is not an entirely new trend. For many years there has been a creeping tendency for consumers to have to do more and more work to 'service' themselves and their families.

One obvious example is in shops. From a situation where paid shop-workers did everything from weighing and wrapping goods, to getting them down from the shelf, putting them in a bag and, in many cases, delivering them to the customer, we have moved to one where it is usual for shoppers to select their own goods from the shelves, wheel them around the shop, wait in a queue to pay for them (so it is their unpaid time, not the shop-worker's paid time that is wasted) and finally lug them home. There are parallels in most other services.

In medicine, it used to be the doctor's paid time that was spent travelling to your door with his or her little black bag when you were ill. Now it is generally you who must transport yourself to the clinic and wait until there is time for you to be examined.

Studies of productivity in banks in the 1960s showed that the greatest single increase in productivity they could achieve, using existing technology, was by persuading customers to fill out their own deposit slips.

The introduction of new technology across the service sector has opened up many more possibilities for unloading this sort of work onto consumers, as more and more of the services we are used to become automated. Door-to-door postal deliveries, for instance, could soon become a thing of the past. Those who could afford it would have gadgetry in their living rooms for receiving electronic letters. Others would have to walk to the nearest public receiving station, just as today they have to go out to poorly functioning public call-boxes and launderettes, if private technology is beyond their means.

The process by which these new forms of self-service get taken on is quite interesting, because they are usually welcomed by consumers, at least in their early stages.

When a new bit of self-service technology, say a launderette, or a cash-dispensing machine, first comes into use, it is in direct competition with an existing service (in this case, a laundry, or a bank counter). It is almost invariably cheaper and more convenient. The brand-new machine does not go wrong, and is quick to use. You can avoid waiting in a queue or hanging around while someone else performs the task for you. It will also be cheaper, because the owner is not having to pay for labour to operate it. As the machine comes into more general use, fewer and fewer people use the old-fashioned service, which has to put its prices up and become even more expensive to keep going. Meanwhile, the machines are being mass-produced and becoming cheaper and cheaper. So even more people use them, although there may be many, such as the old, the illiterate or the disabled who for one reason or another find them impossible. Eventually, as this process continues, the traditional service dies out altogether; there is no longer any choice; you use the machine or you do without. This is often the point where the problems start appearing. Then if the machines become increasingly expensive to use, there is nothing you can do about it, apart from trying to do without the service altogether.

There is also little you can do when they go wrong, swallow up your money without performing, or otherwise frustrate you. You just have to keep your temper and find another one. There isn't a person around anymore who can help.

Obviously, this trend will not affect all services equally or at the same time, and some of the disadvantages of the new machines will be counterbalanced by advantages. Nevertheless, it is fair to say that the trend towards an increasingly self-service economy is a well established one, and it is difficult to see any developments which might reverse this trend, a trend which means that ever-increasing amounts of 'self-service work' will be foisted onto consumers.

Now how, you may be wondering, does this development particularly affect women? We know that women are more likely to work in the services, but surely everyone uses them? The answer is that although men consume quite as much as women do, women do much more of the *consumption work*.

This seems to be mainly because it is so closely associated with housework and housework, in turn, is seen as exclusively female work. Indeed, the words 'housewife' and 'consumer' are often used interchangeably, especially by politicians around election times, when they make speeches based on the assumption that women, as housewives/consumers are exclusively concerned about prices (whereas men/workers are concerned about wages).

Women's low wages, and the fact that they have always done a great deal of unpaid housework in the past, can lead to increasing amounts of the new types of consumption work being foisted onto them, even when it may be a type of work normally done by men in the past. This is because it is assumed that a woman's time is less valuable than a man's (which is, of course, true when he is earning higher wages) so it matters less when her time is wasted. So women may find themselves adding to their traditional household chores such tasks as queueing at the garage to fill the car up with petrol when there's a shortage, waiting in for the plumber, electricity, gas or other repair people, or traipsing down to the offices to do battle with their superiors when they don't show up, or taking on increasing amounts of do-it-yourself repairs or decoration around the home.

So, women do most of the family shopping and routine consumption work. They are also more likely to be users of social and health services since not only do women themselves need to visit doctors and clinics much more often than men because of the medicalisation of contraception and childbirth, but they are also much more likely to be the ones to accompany children, the old and the disabled when they need treatment.

All these chores add a considerable amount of time—and a great deal of frustration and boredom—to the average woman's working day. At a time of deteriorating transport, housing and other public services, these demands can only become greater, adding more stress and more unpaid work to women's lives.

9

New Technology in the Home

One of the most alluring aspects of the new technology is the promise it holds out to women of bringing liberation by automating housework practically out of existence.

If new technology really could take over the cleaning, the cooking, the washing and all the other household chores that keep housewives trapped all day in their homes, many women believe it could release them to go out and do more satisfying work, earn some money or just enjoy themselves. For them these advantages may seem so great that they more than outweigh the disadvantages of new technology they may be aware of in the workplace.

How much truth is there in this idea? Does new technology in the home really shorten the hours of housework and free women to do other things?

If it does, then we might expect the amount of time spent on housework to have been going down steadily ever since labour-saving devices started being introduced into the home. It is a puzzling fact, however, that it hasn't. What little research there has been shows that, if anything, the average number of hours of housework has been going *up*, from around 60 hours a week before the second world war to over 70 in the sixties and seventies.

How can this be explained? One clue comes from the manufacturers of labor-saving equipment themselves. Look at this advertisement for a microwave oven.

This carries a lot of different messages to the woman reading it, some of them quite contradictory. It promises her, for instance, that she will be liberated, that she will be freed from 'slaving over a

Nothing copes with three sittings a night quite like a Sanyo.

The comings and goings of the average family can cause quite a few problems as far as feeding times are concerned.

They all want to eat when it suits them. Not the cook.

So if mum isn't as organised as a catering corps chef, she's in trouble.

Unless she's got a Sanyo Microwave Oven.

With a Sanyo she can cook meals till the cows come home, never mind the rest of the family.

Apart from being astonishingly quicker than the old cooking methods, it's cleaner, more convenient and more economical, as well as being healthier by retaining all the nutritional value of the food.

Preparing three or four different meals a night is simply a pleasure with Sanyo.

Everyone can eat what they want, when they want, with perfect results everytime.

You can choose from five different Sanyos. They'll all prepare meat and two veg in the time it would take to fix beans on toast the old way.

One model incorporates a temperature probe which will automatically shut off the cooker when the food has reached a pre-set temperature. Another has a unique hot-air convector browning fan to make your food look as mouth-watering as it tastes.

A Sanyo is the only way to cook. The advantages are too numerous to mention on this page.

So fill in the coupon and we'll send you all you need to know.

And when you're in the kitchen, slaving over a hot stove, think of the Sanyo and what you could be doing.

– – – – – – – – – – – – – – –

Please send details of Sanyo range to

Name _____

Address _____

RDB2

Jane Bates, Sanyo Marubeni U.K. Ltd.,
8 Greycaine Road, Watford WD2 4UQ.

SANYO
Technology at your fingertips

– – – – – – – – – – – – – – –

Picture by courtesy of Sanyo

hot stove', and implies that the microwave oven is quick and easy to use. However it doesn't say that it's so easy to use that maybe Dad and the kids could prepare their own meals. No, cooking remains very firmly Mum's responsibility. Even more insidiously, the ad actually gives her *more* responsibility than women had in the past. Traditionally, a woman cooked one meal in the evening, and members of the family knew when that was going to be served up. If they couldn't be there at that time to eat it, then that was their hard luck. They had to eat something cold instead or warm it up themselves, or it would be left waiting in the oven for them, while Mum got on with something else. Now it seems, she is expected to cook *three* meals a night, and to hang around at the beck and call of all the members of her family, ready to produce these hot meals the instant they want them.

While it may well be true to say that the microwave oven makes cooking any given meal quicker than it was in the past, there is no guarantee that the time saved will be used by the woman for other things. If she behaves as the advertisers would like her to, she cooks more often, because expectations and standards have gone up.

A similar process has already taken place with many of the other tasks normally done by women in the home. Synthetic easy-care fabrics and washing machines have merely meant, in many cases, that people now expect much cleaner clothes. In households with young children it is now usual for washing to be done daily, when it used to be done weekly, fortnightly or even monthly. Most of us expect clean underwear every day, although our great-grandparents may have been sewn into theirs for the winter, not to be unstitched till the spring. The 'spring clean' was often, too, the only thorough cleaning many houses ever had. Now, the gleaming plasticised surfaces in every corner of our homes are a source of shame if they are not kept constantly shining with the super-duper cleaning products advertised in the women's magazines and on TV.

So, it seems that the new appliances and products that science and technology have brought into our homes, when it comes to the amount of work that has to be done, have taken with one hand while they have given with the other. They have enabled standards to be pushed up higher than could have been conceived possible a hundred years ago but this has created additional work. Of course technology, in itself, has not pushed standards up in this way. Women have been educated to use it in certain ways by advertisements, by the media, and by various forms of education such as domestic science classes at school and handouts in clinics.

This raising of standards is one of the reasons why housework still takes up such a huge proportion of most women's time. However it is not the only one. A second reason is described in the last chapter – the way in which automation of 'service' jobs thrusts more and more 'self-service' work onto consumers, so that new, time-consuming tasks are constantly being added to housework.

A third reason is that in a small household there is no economy of scale. When new technology is brought in, in a factory, a shop, an office or some other workplace, it saves time and labour usually because it allows a task to be repeated over and over again without much effort in one continuous flow. The saving comes when a large number of items have to be processed. For instance, it makes sense to use an addressing machine if you have lots of letters to send, or an electric whisk if you have to beat enough eggs to make 200 school dinners. In the home, most tasks are done on a very small scale, and for a very short period of time, for instance most of us rarely cook for more than three or four people, or clean more than two or three rooms at a time. Yet the time taken to get a machine out, set it up, clean it, dismantle it and put it away again are exactly the same whether we are using it for two minutes or two hours. While people continue to live in small family units, each duplicating the many different tasks all the other family units are doing, on a tiny scale, then new technology can never bring the sort of savings to the household that it can bring to a big workplace.

There is yet another reason that the scientific and technological developments of recent years have not reduced the amount of time women spend on housework. This is connected with the fact that part of the housewife's role is caring. It is the woman's responsibility to keep her children, and any sick or elderly dependents, *safe*. TV road safety advertisements and posters in clinics carry the same message – women should be constantly vigilant. Take your eye off that toddler for one instant; leave a saucepan on the stove while you answer the phone; forget to look in on granny before she goes to sleep; leave your valium on the bedside table or the bleach bottle on the bathroom floor – and you could be responsible for a dreadful accident. There is never any doubt that it is your fault, in the minds of the copywriters who concoct these advertisements, and most mothers are haunted by the fear that one day they will cause such a disaster. Many hours of the time spent in the home are in fact taken up with what one might call 'safety work': taking the precautions that keep the environment safe, and simply guarding. As anyone who has had the responsibility for a lively toddler knows, any task takes three times as long if you have to keep one eye out for the child's safety all the time.

But what, you may ask, does this have to do with science and technology? Surely they have made the world safer than it's ever been before? In fact, the reality is a good deal more complicated than this. Home safety seems to be another area where 'progress' gives with one hand and takes away with the other. On the plus side, many of the hazards our forebears had to face, such as open fires, dangerously poor sanitation and extremes of temperature, have been removed. Gone too are many of the heavy, dirty chores, which veined the legs and roughened the hands and bent the backs of our great-grandmothers. But in their place we have a bewildering range of new hazards. Each of the important scientific and technological advances of the past century seems to have brought with it its own particular array of horrors.

The internal combustion engine, for instance, has made the environment surrounding most homes a lethal one for children and adults who are not fit and alert. It is estimated that one child in four will, at some time, be involved in a traffic accident. No longer can any child be left safely to play out of doors, as was the norm in most parts of this country until fifty years ago – a development which has added enormously to the supervisory work of parents, and created new stresses for both parents and children, who are now deprived of any break from each other's company.

The chemical industry, which has brought us many valuable drugs and almost miraculously effective labour-saving products for the home, has also introduced a baffling variety of new poisons, the effects of many of which are not even fully known. Look at any bathroom cabinet, or the shelf under the sink where cleaning products are kept, or the cupboard with the glues and the do-it-yourself products and read some of the labels to get the idea. There are literally hundreds of products which 'must be kept away from children', 'must not be used near pets', 'must not be sprayed near food', 'must not be inhaled', 'must not be used near a naked flame',

Whose ideal home?

For the vast majority of
housewives, new
technology will not
magically bring a
gleaming new home.
When it is introduced,
it will be in the form of
a new gadget or two, for
which space will have to
be found in the existing
accommodation, with all
its inadequate design
and dangerous clutter.

Picture: Margaret Murray

'must not be punctured or incinerated', 'must be kept well away from the eyes', or bear the instruction, 'call a doctor immediately if accidentally swallowed.' Why? Because they're all lethal. And, unlike most of the hazards of the past, few of us even know how or why they're lethal, so we cannot take remedial action ourselves, but must trust blindly in the 'experts' and pray that they know what they are doing. To judge from the number of new hazards constantly coming to light (well-known products suddenly taken off the market, or appearing with new cautionary labels) it seems probable that many of them don't, and we may be exposing ourselves to hazards we don't even know about, as well as expending time and energy on protecting ourselves and our families from the ones that are known.

Electricity is another example of a major technological advance that immeasurably raised the standard of living of all but the richest when it was introduced into people's homes. Evening activities needing good lighting, new forms of entertainment, new ways of storing and cooking food and a huge range of different appliances all became possible with its introduction. But . . . It also brought new fire risks and turned most rooms in the average house into places where a small child cannot be left alone, for fear of little fingers getting caught in whirring motors or stuck in sockets; of heaters or cutting tools or sewing machines being accidentally switched on; or of television or hi-fi sets brought crashing to the floor as a result of a foot caught in a trailing lead.

There seems to be every likelihood that other forms of new technology will have similar contradictory effects on the working lives of housewives, bringing a mixture of new freedoms and new hazards in their wake.

Even when it comes to the effects of technology on the skills and quality of the working life of the housewife, we find the same mixture of gains and losses.

In the past, housework skills were often quite difficult to acquire, but, once acquired, they gave their owner a certain amount of respect and status. In small communities, women became famous for their beautiful darning, their delicious sponge-cakes, or their well-stocked pantry shelves. This meant that they were well and truly trapped in their roles as housewives, since husbands and others could not be expected to learn these difficult skills and share them. On the other hand, they were secure in their roles. A husband would be unlikely to leave a woman who was keeping house for him so well, and her decisions within the house were not challenged. She was the expert.

The present situation is somewhat different. Learning how to do something in the house is generally merely a question of watching the ads on the telly, or reading the instructions that come with the appliance or on the packet. Anyone can do it. The fact that 'anyone' includes men, can mean that in some households housework chores can be shared equally between men and women. More often, though, it simply means that the woman does the work, but no longer gets any satisfaction or prestige from doing it. Husbands who understand how the machines work, and who have read the instructions on the packets feel that they know how these tasks should be done, and become free to criticise and to demand higher standards of housework. Because standards have become so uniform, women as housewives also become much more interchangeable with each other. No longer can a woman feel secure in the knowledge that 'there's no one can get his shirt-collars as white/make his favourite steak and kidney pudding as I can'. For many women the long-term effects of this new technology entering their homes have been a loss of status, an increasing sense of precariousness in their relationships and a loss of the satisfaction and self-respect which comes from being the 'expert' on your work. These days, the 'experts' on housework are the white-coated scientists on the television, and the technicians who – after endless delays – come round to fix the washing machine or the fridge when it has gone wrong.

So far in this chapter, we have looked at technology which is already familiar, which, like it or not, most of us have come to live with and even depend on. What new developments does microprocessor technology have waiting in store for us?

Most of us will have seen, at one time or another, media representations or exhibition mock-ups of the 'home of the future', all-electronic antiseptic-looking spaces in which everything is designed in exactly the same style. At the outset, it is probably worth saying that it is highly unlikely that ordinary people's homes will ever be at all like that. This is because change does not come about overnight, in a clean sweep, destroying everything that has gone before. For the most part, it takes place gradually, in a series of unco-ordinated fits and starts. People acquire new gadgets in a piecemeal way, usually one by one. They probably do not match each other, and will be fitted into a home which already contains a confusing array of things, old and new, reflecting a variety of tastes and interests.

Class differences will not magically disappear, and there will undoubtedly be wide disparities between what the media present as desirable, and what a few rich people actually own, and what most working-class people can afford. As new gadgets spread more widely among the middle classes, having to live without them will become more and more of a nuisance for poorer people, since publicly available services generally deteriorate as private owner-ship grows. Examples of this in the past have been the deterioration of public transport with the growth of private car ownership, and the deterioration of launderettes with the growth of washing machine ownership.

So, what are these new developments likely to be?

Firstly, there will undoubtedly be a continued increase in the sophistication and complexity of existing appliances, as virtually all electrical appliances (food mixers, clock radios, central heating control systems, cookers, vacuum cleaners, irons and so on) incorporate programmable microprocessor 'brains' enabling them to switch on and off automatically and perform a wider range of functions than their traditional counterparts.

There is also likely to be an increase in the range and versatility of home entertainment systems, with a proliferation of gadgets that can be attached to your television screen to give it new functions: video games, video recorders and playback systems and information systems such as ceefax, oracle and prestel. For those of us who cannot afford such systems, the indirect effect will probably be still further deterioration in the quality and choice of standard broadcast television and radio programmes, and in out-of-the-home entertainment like the cinema.

Computers will become more common in people's homes too, carrying out a range of functions from helping kids with their homework (yet another new way in which class divisions may be perpetuated – the children whose parents cannot afford a computer will be at a disadvantage at school) to working out accounts, or storing information.

Household robots are also a real possibility. There is already technology capable of producing machines to carry out a wide

The listening, talking oven

New technology brings increasing sophistication to traditional domestic products. This 'Show and Talk' microwave oven from Japan, for instance, can perform a variety of new functions: it can be operated almost entirely by verbal instructions. Using voice recognition techniques, for example, the door can be opened simply by saying 'open door'. It can also talk back to the user, using voice synthesis, telling you things like how much cooking time is left, or confirming your instructions. Another feature is the 6-inch television monitor which can show you menu selections, ingredients, cooking instructions, or keep you up-to-date on power consumption or cooking progress. When not engaged thus, it can receive broadcast TV programmes or, with a closed-circuit camera attached, monitor what is going on in other rooms in the house. Finally, in case you want to record any of its information for posterity, the machine can print it out for you with its own built-in printer.

Picture by courtesy of National Panasonic

variety of such household functions as scrubbing, sweeping, polishing and vacuum cleaning. One such machine, described in the *Guardian* in 1978, comes 'complete with a skirt' and 'made small so as not to be intimidating'. However it seems more likely, at least in the shorter term, that these functions will be carried out by increasingly sophisticated versions of existing appliances, as described above, rather than by all-purpose robots.

Other changes in the quality of life in the home will come about as a result of technological developments outside the home, such as the growth in fast-food and in vending machines (which can lead to an increase in takeaway and convenience foods), the development of electronic mail and changes in shopping patterns. The growth of direct ordering of shopping from the home, either electronically through systems like prestel or by mail is almost certain to increase.

In the longer term, the development of ever more sensitive 'feeling' technology seems likely to lead to even more far-ranging changes in the home. At the moment, home-based applications of this type of equipment are almost exclusively limited to surveillance systems, burglar alarms and smoke-detection. However it could be developed for many other purposes. One commentator, writing in *Now* magazine in 1980, prophesied that by the end of the decade we would have 'the cot computer' which could note a baby's breathing pattern, analyse its voice and observe its responses so that it could tell when a feed should be provided, when a nappy needed changing and so on. Not only could it predict these needs, but it could meet them, by providing thermostatically controlled milk at the right temperature or a furry cuddle when required.

To attempt to predict further than this is to indulge in idle speculation. But, if we can learn anything from what has happened when technology has been brought into our homes in the past, it is that the advantages it brings will be mixed with disadvantages.

It seems unlikely that new technology will liberate women from housework until the day comes when housework is no longer seen as a woman's responsibility. It may become easier, it may change its nature out of all recognition, but while it's still considered to be menial work, and women's work, women will continue to be trapped by it.

10

Women, Unions and New Technology

The public image of a trade unionist is still usually that of a man, a 'brother' in grubby overalls who uses incomprehensible bureaucratic language, goes on strike at the drop of a cloth cap and devotes his spare time to 'holding the country to ransom'.

Needless to say, this image does not bear very much relation to reality: four out of ten trade unionists are now white-collar workers; at least a quarter vote Conservative; two thirds have never been on strike; and a third are women.

It is this last fact, the high proportion of female members, which you could most easily be forgiven for not knowing. Despite a record of militancy at least as good as men's when given a chance, these women, who have been joining unions at twice the rate of men over the past decade, have been all but invisible to the public eye.

When it comes to representation for women in their unions, it hardly exists.

This table shows the numbers of women in ten of the unions with the largest female membership in this country, together with the numbers of women representatives. The figures in brackets show the numbers of female representatives there should be if women were represented according to their share of the membership. As you can see, the disparity is enormous. Even in the best example, NALGO, which has fourteen women on its executive, this is still less than half the number it should be, if representation were proportional.

The reasons for this state of affairs are depressingly familiar: women are discouraged by the sexist attitudes of male officials;

Women in the unions

Figures in brackets show how many women there would be if they were represented according to their share of the membership.

Union	Membership			Executive members		Full time officials		TUC delegates	
	Total	F	%F	Total	F	Total	F	Total	F
APEX (Professional, Executive, Clerical, Computer)	150,000	77,000	51%	15	1(8)	55	2(28)	15	4(8)
ASTMS (Technical, Managerial)	472,000	82,000	17%	24	2(4)	63	6(11)	30	3(5)
BIFU (Banking, Insurance Finance)	132,000	64,000	49%	27	3(13)	41	6(20)	20	3(10)
GMWU (General & Municipal)	956,000	327,000	34%	40	0(14)	243	13(83)	73	3(25)
NALGO (Local Govt Officers)	705,000	356,000	50%	70	14(35)	165	11(83)	72	15(36)
NUPE (Public Employees)	700,000	470,000	67%	26	8(17)	150	7(101)	32	10(22)
NUT (Teachers)	258,000	170,000	66%	44	4(29)	110	17(73)	36	7(24)
NUTGW (Tailor & Garment)	117,000	108,000	92%	15	5(14)	47	9(43)	17	7(16)
TGWU (Transport & General)	2,070,000	330,000	16%	39	0(6)	600	6(96)	85	6(14)
USDAW (Shop, Distributive Allied)	462,000	281,000	63%	16	3(10)	162	13(102)	38	8(24)
Totals	6,022,000	2,265,000	38%	316	40(150)	1,636	90(640)	418	66(174)

All figures are approximate, and the most recent that were available in November 1980. Taken from *Hear this, brother*, by Anna Coote and Peter Kellner, New Statesman Reports No 1,1980

women cannot attend meetings because of family commitments; women are concentrated in the workplaces where union organisation is weakest; women have been deprived of the chance to acquire the skills that would enable them to break through into the well-paid, well-organised craft areas; and so on. Perhaps the most important single factor is the simple fact that women have children, and are still expected to take responsibility for bringing them up, looking after the household and servicing men; they are defined primarily as housewives and their paid work is seen as a dispensable extra, not an important and central part of their lives.

As a result of the agitation of women in their unions over the last decade or so, a glimmering awareness of some of these problems has begun to emerge, and several individual unions and the TUC have adopted policies of seeking equality and working towards improving the situation of their women members.

The TUC's Charter, *Equality for women within trade unions*, is one of the most important of these, and is reproduced here to show the sorts of demands that can be raised in your union.

At the 1981 Trades Union Congress, a more detailed policy, outlining *positive action programmes*, was passed. Copies of this can be obtained by writing to the TUC at Congress House, Great Russell Street, London WC1.

At present, these remain paper policies; they will only be translated into action if they are taken up by large numbers of trade unionists in their own workplaces and union branches.

When it comes to new technology, a similar picture emerges. Since 1978 a large number of unions, especially those with a high proportion of white-collar workers, have busied themselves with special committees and working parties to look at the effects of new technology on their members. To their credit, several have realised that the effects are likely to fall much more heavily on women than on men, and have made special attempts to highlight these problems. Again, though, too many of these policies have remained on paper, and it has been only in exceptional cases that they have been translated into action.

One of these exceptional cases was in Hull, in early 1981, when 1,000 women members of the Transport and General Workers' Union at Reckitt and Colman's went on strike to defend their right not to work with new machinery without negotiation and agreement. They had been trying to negotiate a new technology

TUC Charter

Equality for women within trade unions

Commended by the General Council of the Trades Union Congress 'to all union executives and committees . . . for the integration of women within trade unions at all levels'. Endorsed by Congress, 1979.

1 The National Executive Committee of the union should publicly declare to all its members the commitment of the union to involving women members in the activities of the union at all levels.

2 The structure of the union should be examined to see whether it prevents women from reaching the decision-making bodies.

3 Where there are large women's memberships but no women on the decision-making bodies special provision should be made to ensure that women's views are represented, either through the creation of additional seats or by co-option.

4 The National Executive Committee of each union should consider the desirability of setting up advisory committees within its constitutional machinery to ensure that the special interests of its women members are protected.

5 Similar committees at regional, divisional, and district level could also assist by encouraging the active involvement of women in the general activities of the union.

6 Efforts should be made to include in collective agreements provision for time off without loss of pay to attend branch meetings during working hours where that is practicable.

7 Where it is not practicable to hold meetings during working hours every effort should be made to provide child-care facilities for use by either parent.

8 Child-care facilities, for use by either parent, should be provided at all district, divisional and regional meetings and particularly at the union's annual conference, and for training courses organised by the union.

9 Although it may be open to any members of either sex to go to union training courses, special encouragement should be given to women to attend.

10 The content of journals and other union publications should be presented in non-sexist terms.

agreement for some time with an intransigent management that finally installed a new packaging machine without agreement, and suspended the 30 women who refused to operate it in accordance with their union policy. Despite financial hardship and blacklegging on the part of the men in the factory, whose maintenance jobs were not at risk from the new machines, they stayed out for a month, and returned to work undefeated. There are few other instances where such determined action has been taken over new technology.

Perhaps the most celebrated example of a new technology dispute was the long strike at Times Newspapers in 1979–80 to defend jobs and insist on the unions having a full say in the terms on which new technology was introduced. Without wishing to denigrate the strength and determination of the trade unionists there, it is worth noting that they were in a position few other workers in manufacturing could match. If you work in a factory making TV sets, or car parts, or chocolates, or shoes, and decide to go on strike, it is an easy matter to import substitutes from abroad for what you make, to undermine the success of your action. But the *Times* management could not have imported *Le Monde*, or the *Washington Post* or another foreign newspaper to substitute for the non-appearing *Times*; they simply would not have been relevant to the newspaper buyers. They did in fact make an attempt to get newspapers printed in Germany during the dispute (and were foiled by the solidarity of German workers) but it would still have required English-language journalists, based in Britain, to write the stories.

In addition, of course, when a day's newspaper production is lost it is lost for good. It cannot be sold on the following day, or the day after because it has gone out of date. These factors put newspaper workers in a much stronger position than most other workers in manufacturing industry when it comes to negotiating over new technology.

Nevertheless, there is still a great deal that can be achieved by groups of workers seeking, if not to prevent new technology being brought in altogether, at least to ensure that workers do not bear the entire social cost when it is introduced, and to have some say in the terms and conditions of its introduction. One lesson to be learned from the experience of the past few years is that if this is to take place, then it is up to the workers at that particular workplace to make it happen. Trade union officials can draw up policies, give advice and support, and carry out research, but it is only the members who can take the action that will make a reality of these policies.

The next section of this book is a checklist of points to consider when embarking on negotiations over new technology.

11

Negotiation Over New Technology

As we have seen, the effects of new technology vary considerably between different types of workplaces. There are also wide differences in the degree of trade union organisation, and the bargaining strength of particular unions from one workplace to another. Probably no two groups of workers faced with the introduction of new technology will come up with exactly the same set of demands, and what it is possible to win will depend on a number of factors – how well organised and determined the union is, how well the employer is doing, the scarcity of the workers' skills and so on.

For these reasons, this section of the book is not intended to be a blueprint for how to negotiate the perfect new technology agreement (if, indeed, there is such a thing); it is meant as a checklist of points that may be relevant in your workplace, that could serve as a starting-point for discussion with your fellow trade unionists about how you should be responding to the introduction of new technology. If some of these demands seem like asking for the moon in your workplace, remember that they may seem quite reasonable somewhere else.

Whatever set of demands you finally come up with, the most important thing is that they should be what the workforce wants, and what it is prepared to fight for. There is no point in going in to your management with a long list of demands that most of the members don't understand or want, and that management is perfectly aware no one will take action over. You will be laughed out of the negotiating room and achieve nothing.

Importance of information

The importance of being well-informed about the new technology before you begin finalising your claim cannot be over-emphasised. Most trade unions have produced helpful literature for their members on this subject, and you could start by asking your full-time official for copies, or writing off to the union's head office. In addition, many unions have research departments, which will provide additional information on request. Don't be afraid to demand help of this sort from your union. After all, it's what you pay your subscriptions for.

As well as information from your union, you should be able to get information from your management about any plans they have for introducing new technology, the reasons for this (including information about productivity, staffing levels etc.) and the type of equipment they are proposing to introduce. If your union agreement already has a disclosure-of-information clause, this would be the appropriate procedure to use. If not, you will have to request it separately. Remember that, under the Employment Protection (Consolidated) Act, recognised trade unions have a legal right to information from the employer which is relevant to their collective bargaining.

Having the right to this information, however, is a different matter from getting it, and you may have to face a tough fight to extract all the information you require from a reluctant management, who may try to exploit every possible loophole in the Act to avoid telling you what you want to know.

If you find that you require still more information, you could try sending your shop steward or safety rep on a new technology course run by your union or by the TUC. Details of the TUC 5-day new technology courses can be obtained from your regional TUC education officer, or by writing to the TUC Education Department at Congress House, Great Russell Street, London WC1. Failing that, you may be able to find relevant literature in your local public library, or from trade journals. If there is a trade union resource centre in your area, the staff there may be able to help with information, contacts and advice.

If your organisation is a large one, another source of information might be other trade unions represented there, or unions in other branches of the business. If there is a shop stewards' committee, or combine committee or some other form of joint union committee, this would be the place to start exchanging this type of information. If there isn't such a committee, then maybe it's time you considered setting one up. One of the effects of new technology is to break down the traditional differences and demarcations between different groups of workers, so it's more important than ever for unions to work together over its introduction if they're to avoid a divide-and-rule situation.

Again, this is much easier said than done, especially during a recession. Workers in different parts of the organisation in different parts of the country, or even different departments on the same premises will often feel that they are in competition for jobs. A shop steward in one section may very well believe that the most important priority is to safeguard jobs in that particular section, even if that means fellow-workers in another part of the organisation going to the wall. Developing a broader perspective and a spirit of solidarity is no easy task. Demoralisation as a result of past losses or defeats can make this even more difficult.

Importance of involving all the affected members

As well as establishing the best links you can with the other union branches, it is important that you involve as many as possible of the workers who are likely to be affected by the introduction of new technology. If they are not in the union, encourage them to join. If they are, try to ensure that your meetings are held at such times and places that everyone can attend. Ideally, meetings should be held during working hours; if this is not possible, maybe it is possible to organise them at times when women do not have too many other pressing obligations, and to arrange some sort of childcare during the meeting. Don't forget the part-time workers and their needs when working out the best time for your meetings.

It is a good idea to prepare for the first meeting by distributing literature explaining what it is about. The literature provided by your union may be appropriate. If not, perhaps you could prepare a duplicated leaflet outlining the main points. If the subject is a new one to all of you, it might be an idea to invite someone along to introduce it, perhaps a worker from another

similar workplace where new technology has already been introduced, or someone from your union office. In some parts of the
country, trade union resource centres or women's groups are able to
provide speakers for meetings of this sort.

Drawing up your demands

Here are some of the points you may find it useful to include in your
new technology claim:

Disclosure of information

**Management should provide the union with full informa-
tion, as far in advance as possible, about its investment
plans, including specifications of any new equipment or
systems it intends to introduce, together with information
about proposed future staffing plans.**

Note: Wherever 'the union' is referred to in this check list it
means the union representatives at the workplace, not the full-time
official from the district office. Naturally, these representatives
should not consent to anything without first securing the agreement
of their members.

You may wish to tighten up this clause by specifying a
definite time, e.g. six months or one year, instead of the vaguer 'as
far in advance as possible'.

Procedure

**A procedure for consultation over these plans should be laid
down, with a clause specifying that no new equipment or
systems shall be introduced without the consent of the
union.**

Note: In all discussions over advance plans for introducing
new technology, as with other changes in work procedures, the onus
should be on management to prove to the union that it is necessary,
not on the union to resist it. You should start from the assumption
that the status quo prevails, and that any changes which might
adversely affect union members should only be introduced if
management is prepared to pay the full 'price' for them – in terms of
extra pay, improved conditions, job security etc.

Job security

No redundancies should be created as a result of the introduction of any new technology. The jobs of all existing staff should be guaranteed for a given period of time, and a procedure established for consultation and agreement with the union over future staffing levels when this period runs out.

Note: The details of this particular clause will have to be very carefully worked out in the light of your particular circumstances. For instance you may wish to specify separate staffing levels for particular departments or groups of workers. This may be a very tough or even an impossible clause to negotiate. Before conceding redundancies, you should explore every possible alternative, such as shorter working hours, overtime bans and so on. If you do end up negotiating a redundancy agreement, remember the discriminatory effects of principles like 'part-timers first' and 'last in, first out' and look for an agreement that ensures that management pays the highest possible price for the jobs it is destroying.

Working hours

Working hours should be shortened by a specified amount for all staff with no loss of pay or other benefits.

Note: It is important to be specific about what sort of shorter hours you want, and to ensure that this is what the majority of workers want. In one factory where $2\frac{1}{2}$ hours off the working week was successfully negotiated when new technology was brought in, the union members were divided about when they wanted to take the time off – the women wanted to leave half an hour earlier each day, while the men wanted Friday afternoons off. Since more men attended the union meeting, their desires carried the day. Remember too that a shorter working week will not create extra jobs if everybody simply does more overtime when it is introduced. A final point: one of the most effective ways of bringing about a shorter working week in many workplaces is to improve conditions and wages for part-time workers so that part-time working provides a decent standard of living. It then becomes an attractive option for more workers, thus creating more jobs. Job-sharing may provide an alternative way of bringing about the same effect, but make sure it isn't just providing your employer with two jobs for the price of one.

Early retirement

The retirement age should be brought forward by a specified amount, with no loss of full pension entitlements.

Holidays

Annual paid holiday entitlement should be increased by a specified amount.

Sick leave/maternity leave/paternity leave/compassionate leave/sabbaticals

All these should be increased by a specified amount.

Note: The last three clauses are all different ways of reducing the amount of time workers spend at the workplace, thus increasing the amount of work available to create new jobs.

Changes in working hours

No changes in working hours should be introduced without the full knowledge and agreement of the union.

Note: Depending on your circumstances, you may wish to be more specific about the details of this clause, which is designed to prevent management introducing shift systems, night working or other changes without your consent.

Homeworking

The employer should not put any work out to homeworkers without the full knowledge and consent of the union. Where homeworkers are employed, the union should be provided with their names and addresses, details of the work they are given, and rates of pay. The union should be given every opportunity to recruit them into membership, and to negotiate on their behalf.

Note: This clause is to ensure that management does not use homeworkers as a form of cheap labour, to undercut union organisation at the workplace. There may well be circumstances in which you wish to agree to the use of homeworking. If so, these are some of the things you should negotiate for on behalf of the homeworkers: pay (which should not be less than that for staff

Workers resist the new technology

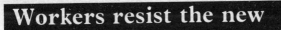

Printers on *The Times* newspaper during their 11-month strike to demand no redundancies and a new technology agreement while (below) the new machines stand idle.

Pictures: Chris Davies/Network and Judah Passow/Network

workers); employee status; expenses; health and safety; union membership.

Grading

No worker should be downgraded as a consequence of the introduction of new technology.

Note: You may wish to take a critical look at the grading structure, in the light of the new technology, and put in a demand for *up*grading of some or all groups of workers whose jobs are affected. Remember that you may have to live with a new grading scheme for a long time, and think of the long-term implications.

Pay

Many unions have negotiated pay increases for workers whose jobs are changed by new technology. You may wish to do the same.

Note: While it is obviously important to get as good wages as possible for your members, and to make management 'pay' for your co-operation with their automation plans, be sure that wage increases do not simply become a way of bribing workers to accept new technology at the expense of jobs and of working conditions. Remember, a wage increase is eroded away in no time by inflation; a lost rest-break or other non-wage advantage is lost for good.

Productivity

No changes in productivity targets or job timings should be set without the full knowledge and consent of the union. All data on productivity collected by the management should be made available to the union.

Note: This clause is designed to prevent speed-up of the work without negotiation, and to enable the union representatives to keep themselves informed of members' productivity rates. In the TUCRIC survey, we came across one workplace where the workers did not believe that there had been any increase in productivity following the introduction of new technology; according to their management, it had gone up by 60 per cent!

Operating the new technology

No worker should be required to operate the new machines or systems against his or her wishes. Where workers do not

wish to operate it, suitable alternative work should be provided at or above their existing grades, with no loss of pay or worsening of conditions of employment. Any necessary training to carry out such alternative work should be provided by management at full pay.

Note: You may wish to add extra details to this clause, depending on your circumstances, e.g. you may wish to specify a trial period during which workers can try out new jobs and decide whether or not they want to continue doing them. It's important to avoid a situation where older workers, or those who are not well adapted physically to the new types of work are penalised, and shoved into degrading second-rate jobs for the rest of their working lives because they cannot adapt to new machinery. Still worse is a situation where they are sacked because they are unsuited to the new work. In the US, one of the printing trade unions managed to negotiate for managements to pay for members in this situation to attend a three-year college course of their choice, to retrain for new occupations. Some print workers came back as journalists as a result.

Training

All workers should be given the opportunity, at management's expense and in work time, to train for promotion within the organisation.

Note: It is important that people are not just trained to do their existing job, or to operate the particular new machine they will be required to use, but are given a chance to train for better jobs within the organisation. It is worth remembering that the Sex Discrimination Act allows for women to be given special training to enable them to apply for jobs that traditionally have been held by men. You may wish to specify details of the types of courses you wish to be provided; e.g. in-company courses, or courses run by equipment manufacturers may be very narrow, only training you to work one particular type of machine, and you may prefer a course independently run, perhaps by a local college, which gives you a broader view, and does not have a vested interest in one particular way of working. Some unions also like to insist that longer courses include an element of trade union education, or at least that the union is involved in designing the course.

Machine monitoring

No new machine or work system should be introduced which includes facilities for monitoring workers' work, without the full knowledge and consent of the union.

Note: It can often happen that workers do not even realise that their working speed, the number of breaks or the quality of their work are being monitored by the machines they are working on. Sometimes a machine has this facility, even if it is not used initially. In Denmark, shopworkers went on strike to prevent checkout machines being introduced in one chain of supermarkets which could monitor the operators' work. They won the right for the trade union to appoint a special new technology representative, with the right to inspect all computer programmes to see that they did not include this provision.

Pace of work

All machines and programmes should be designed so that the operator is in control of her own pace of work, to avoid the stress which comes from machine-pacing.

Note: Like the demand for no machine monitoring and some of the health and safety demands outlined below, this may involve close involvement by union representatives in the design of programmes for the machines.

Variety of work

It should be management's intention, in consultation with the union, to work towards a situation where individual workers have as much variety as possible in their daily work. A joint working party should be set up to revise job descriptions with this in mind.

Note: Much of the increased stress resulting from the introduction of new technology comes from the monotony of doing one repetitive task for long periods. In many workplaces, it should be possible to work out some rotation of tasks to avoid this. It is particularly important where some types of work, e.g. VDU operation, carry particular hazards, that operators do not have to do them continuously. Working out the details will vary from one workplace to another. It can only be done effectively with the full and active involvement of the workers concerned. A by-product of achieving greater variety in the work is to improve the promotion prospects of the workers concerned, and their chances of getting alternative jobs in other organisations.

It is very important, however, that no changes in job description should be imposed on workers against their wishes. This demand should only be made with the full knowledge and support of the people who will be affected by it. In many cases, individual workers will object to having to learn new skills, and see an increased variety of work as a threat to traditional demarcation between jobs.

Improvements in workplace facilities

These could include better canteen facilities, a creche, rest-rooms, sports facilities, or anything else your members might want.

Note: This is one way of making sure that your management pays the full price for the extra productivity it is getting by introducing the new technology.

Improvements in workplace environment

Again, these could include almost anything that seems appropriate from carpets and curtains to heating or air-conditioning systems, new furniture or re-decoration.

Note: Safety reps should be fully involved in working out any demands connected with changes in the workplace environment, since many factors affecting health and safety will be involved.

The technology itself

The union should be fully involved in deciding on the design and type of new machinery to be installed.

Note: Here again, there will be many factors involved affecting health and safety. See below for a list of some of the points to look out for. It is in this type of negotiation that you will need to draw most heavily on the sources of information described at the beginning of this chapter. The aim is not to become a complete 'expert' – that would just be doing management's job for them – but to understand enough about what's going on to make sure that the workers' requirements are being met. Don't allow yourselves to be blinded with science, and insist on explanations of anything you don't understand. Generally speaking, if someone is considered intelligent enough to operate a machine then she should be regarded as intelligent enough to understand how it works, at least in its main essentials. And the operator's needs should be the most important in the design of a machine. If, for instance, you don't want to work a machine that requires you continually to stretch your right arm, then this should be something the union demands. It is not the union's job to redesign the machine to make sure that it doesn't happen, but the union representative should be sufficiently clued up to see that this is one of the points included in the designer's briefing, or that there is a chance to test the machine out for this before the union accepts its introduction.

Health and safety

Note: Negotiating over the health and safety of workers working with new machinery is a complex task; demands will vary enormously from one workplace to another, depending on the type of machinery introduced, and the environment of the workplace. For example, the hazards of a sweet factory will be very different from those of a department store, an engineering works, or a hospital. There is no space here to go into the complexities of all these different types of workplace. The points that follow are a summary of some of the main things to watch out for when negotiating over the introduction of computerised equipment using visual display units (VDUs) or microfiche readers, selected because their introduction affects such enormous numbers of women workers. (For other types of new technology, you will need to consult your union for advice.) If you are going to be negotiating over their introduction, you would be well advised to do some further reading, to take account of the detailed specifications of particular types of machines, designed for particular purposes. An excellent book to read on this subject is the *Office Workers' Survival Handbook*, available from the Women and Work Hazards Group, BSSRS, 9 Poland Street, London W1, which has a practical and informative section on negotiating the health and safety aspects of new technology.

● **Rest breaks**

There should be specified maximum lengths of time which an operator may spend at a screen, and a specified break-length. This is because it is important for the eye, which has been focussing on a screen at fairly close quarters, to be able to relax completely, by looking into a distance, and into relative darkness. Different authorities recommend different lengths of break. At the GPO, CPSA members have agreed a code of practice whereby normal time spent with VDUs is a maximum of 100 minutes a day, rising to 180 minutes during abnormal work peaks; no operator is required to work more than 50 minutes without a break of at least 15 minutes. ASTMS members at the Royal Insurance Group have negotiated a maximum of four hours a day.

● **Temperature**
Machines can give off a lot of heat. Agreements should ensure that ventilation is adequate to keep the room at a comfortable temperature.

● **Static electricity**
Carpets and upholstery should be wool or other natural fibres, rather than nylon which can generate static electricity in combination with the machines, causing discomfort to operators, particularly those who are wearing nylon underwear. Some managements have been known to try to ban nylon knickers, to avoid the problem of static electricity – not because they care about the women's comfort, but because it can set the machines off by mistake! The environment, not the individual, should be adapted to avoid this problem.

● **Accident avoidance**
The environment should be designed so that there are no trailing leads, sharp edges or other factors to cause accidents. All machines should be regularly serviced, and there should be adequate storage space and safe surfaces to put things down so that the tops of the machines and surrounding areas do not become littered with things, which could cause accidents if they fell into the works.

● **Lighting**
Lighting is of crucial importance in avoiding eye-strain. Overhead lighting, particularly fluorescent lighting, should be avoided; so should any arrangement in which windows or bright lights are reflected in the screens, causing glare. Recommended lighting

levels for VDU usage are quite dim – around 150 lux, compared with 500 lux in a normal office. The glare index should not exceed 16. Often an operator is required to work moving her eyes backwards and forwards from a paper script to a screen. In such a situation, the best lighting arrangement is an individual desk light which she can adjust to shine onto the paper without reflecting in the screen.

● **Design of work station**

Seating should be adjustable, as should the height of the desk, the angle and height of the screen and the positioning of the keyboard. The keyboard should be separate from the screen so that it can be independently adjusted. The operator's seat should be positioned so that she can look into a distance of at least 20 feet to rest her eyes at intervals, and is not facing a window or brightly-lit wall unless these are screened. Operators should have a say in the design of their work-stations, and as many features as possible should be adjustable, to allow for variations in size and personal preference, and to give a chance to change position during the course of the day. The distance between seat and desk should allow for pregnancy with comfort.

● **Design of keyboard**

There are a number of ergonomic factors involved in keyboard design which are worth investigating. Perhaps the best way to select the most suitable one is for operators to have a chance to try them out and choose which one they prefer. It seems likely that the cases of tenosynovitis, a painful inflammation of the tendons in the wrist, which are coming to light among word processor operators, are due at least in part to poor keyboard design.

● **Design of screen**

Here there are a very large number of factors affecting the health and comfort of the operator. Again, the golden rule is that as much as possible should be adjustable by the operator, to maximise her control and allow for personal variations. Adjustable features should include: **brightness,**
contrast,
screen size and
angle at the very minimum.

The operator should also be able to choose:
the **colour** of characters and background she finds most restful, and, in the case of microfiche readers and VDUs where characters are 'scrolled' up or down across the screen, be able to control the **speed of scrolling**. Other factors that should be negotiated are:
the **flicker rate** (the higher the better: Don't accept a machine with a 'character refresh rate' of less than 60 hertz – in other words, a machine that flickers less than 60 times a second);
the **character size** (which should be large enough to be read comfortably from a distance)
the **line length** or number of characters per line (this should be fairly low)
the '**dot matrix**' or the number of dots that make up one character (the more, the better, from the point of view of readability)
the **picture layout** (operators should be allowed to choose whichever they find easiest)
screen size (it is better for the eyes to look at a larger screen from a distance than a tiny one from close up).

● **Eye-tests**
Many unions demand eye-tests as a precondition for working with VDUs, and at regular intervals afterwards. This is useful for monitoring the effects of the new technology and can be an important way of catching problems before they become too severe – it is estimated that one third of the population has something wrong with its eyesight. It is very important, however, to make sure that eye-tests aren't just used as a way of 'screening out' workers who are unsuited to this work, with the object of putting them out of a job. It is surely an unacceptable situation when people who are over 45 or those who are short-sighted or astigmatic are effectively declared unemployable, because good eyesight has become a condition of work. The emphasis should be on finding ways of minimising the amount of time people have to spend at the screens, and designing machines that are safe to use by everyone. In the meanwhile, trade unions should insist that suitable alternative work is found for anyone whose eyesight makes them unsuitable for work with VDUs, with no downgrading or loss of pay. All eye-tests should be in work time, and paid for by management, who should also pay for any glasses or medication which may be prescribed as a result of the tests. All test results should be made available to the union.

VDU design

VDU design is a complex matter, and choosing the right one for your workplace is something in which the union should be as fully involved as possible. Any one of the factors in this picture could lead to visual strain and fatigue, so all of them should be either negotiated by the union or under the control of the individual user.

Picture based on information supplied by Olov Ostberg

12

Where Do We Go from Here?

For working women directly affected by the introduction of new technology, trade union organisation seems to hold out the best hope of fending off some of its worst effects.

However it would be silly to pretend that trade union action by itself can solve the many problems that the latest wave of automation brings for women.

Large numbers of women workers will find themselves too weakly organised to be able to achieve much through their unions. For many others, like the thousands working for small firms driven into closure or lay-offs by the competition from big, highly-automated companies, there is little that trade unions can do, even in the unlikely event that the workers are trade union members.

For those who are already unemployed, it is too late for unions to achieve much, and it is difficult to see how unaided trade union action can have much impact on the ways in which new technology affects women outside the workplace: the growing threat to civil liberties; the deteriorating services; and the worsening quality of life for so many women.

When it comes to defending the interests of the working class in this, the first serious recession since the second world war, the trade unions simply do not have the answers. If they had, they would have succeeded in saving more jobs, and keeping wage-rates at a decent level. That is not to say that they do not wring some concessions from the employers which would not otherwise be achieved. They are a great deal better than nothing. But it would be unrealistic for women to look to them entirely for their salvation in the present crisis.

So what can women do?

There is no complete 'right' answer to this question, and the first step towards finding one is to recognise this fact. There are no experts who can tell us what to do; it is up to us to work out what we want, and then develop strategies for achieving these demands.

The first step is to come together, in our women's groups, community groups, trade union branches and political organisations, to pool the knowledge and experience we have, and try to build up a picture of the way our lives are likely to change. In some parts of the country, this process has already begun, with groups of women writing articles, running courses and conferences, giving talks and practical demonstrations of computer equipment, to share and spread information about the new technology. In unions, many have agitated for research, training courses and non-discriminatory policies to come to grips with the problems. For some, it has given a new awareness of the importance of continuing to fight for the existing seven demands of the women's liberation movement:

1. **Equal Pay for Equal Work**
2. **Equal Opportunities and Equal Education**
3. **Free Contraception and Abortion on Demand**
4. **Free Community Controlled Child Care**
5. **Legal and Financial Independance for Women**
6. **An end to Discrimination against Lesbians and the Right to our own Self-Defined Sexuality**
7. **Freedom from intimidation by threat or use of violence or sexual coercion, regardless of marital status, and an end to all laws, assumptions and institutions which perpetuate male dominance and men's aggression towards women**

These demands tackle head-on some of the factors that keep women in their secondary, powerless positions in the labour force, and make them so vulnerable to the inroads of new technology.

Other women have pinned their hopes on positive action programmes, and changes in the equal pay and sex discrimination laws.

For others, it has brought a renewed consciousness of the unfairness of the education and training given to most girls, which keeps them in ignorance of the science and technology that will increasingly come to govern their lives. In order to have any chance of getting decent jobs, or of taking any control of their own lives in the future, women will have to learn about the technology for themselves, and free themselves from blind dependence on the male 'experts'. This has led to agitation for new training courses for women, and an end to discrimination in schools and colleges.

Once women develop some confidence in their ability to understand how the technology works and what it can do, it will become easier to formulate demands which could create positive uses for the new technology. Silicon chips open up a bewildering range of options. Perhaps they could be used to improve social services? to make the streets safe for women at night with new alarm systems? to improve shopping facilities and create new jobs by making it possible for the old and the sick to order from their homes, and have the goods delivered to them? to improve communications for the lonely? to bring about a shorter working week for everyone?

One thing seems certain: the technology is unlikely to be adapted to our needs unless we are in a position to say what our needs are, and to organise to make sure that they are met.

This process is only just starting. There is no national organisation conducting it; no programme or policy defining 'the way forward'; just a lot of individual women, in their different ways, in their different organisations, beginning to put their heads together and think. The chances are, they'll come up with some answers that will help all of us. Certainly, nobody else will.

Maurice Frankel/
Social Audit

Chemical Risk

A Workers' Guide to Chemical Hazards and Data Sheets

This book describes the information safety representatives should be given about chemicals used in the workplace – and explains how this information can be interpreted and used. It explains:

- Safety representatives' rights to information, and what they can do if information is refused.
- Why it is essential to know the chemical composition of products – and how safety representatives can deal with the problem of unidentified 'trade-name' products.
- What information safety representatives need about the hazards of chemicals and the handling of emergencies – and what problems they may face from arbitrary judgements and untested chemicals.
- The basic methods of controlling hazards, and the use and abuse of 'TLV' exposure limits.
- How data sheets should be used – and what can be done to detect and deal with unreliable data sheets.
- What information policies should be adopted by employers.
- How to use published sources of information to investigate chemical hazards.

£1.95

This is the latest in the popular Workers' Handbook series, which includes *The Hazards of Work, Rights at Work, Employment Law Under the Tories, Non-Wage Benefits, Using the Media, Pensions* and *The Trade Union Directory*.

For a free information leaflet write to Pluto Press, Unit 10, Spencer Court, 7 Chalcot Road, London NW1 8LH